# EMBODYING ART

# EMBODYING ART

## HOW WE SEE, THINK, FEEL, AND CREATE

CHIARA CAPPELLETTO

TRANSLATED BY SAMUEL FLECK

Columbia University Press
*New York*

Columbia University Press wishes to express its appreciation for assistance given by the Università degli Studi di Milano Department of Philosophy and the Paris Institute for Advanced Study in the publication of this book.

Columbia University Press
*Publishers Since 1893*
New York Chichester, West Sussex
cup.columbia.edu



Library of Congress Cataloging-in-Publication Data
Names: Cappelletto, Chiara, author. | Fleck, Samuel, translator.
Title: Embodying art : how we see, think, feel, and create /
Chiara Cappelletto ; translated by Samuel Fleck.
Other titles: Neuroestetica. English
Description: New York : Columbia University Press, [2022] | Includes bibliographical references and index.
Identifiers: LCCN 2022005430 (print) | LCCN 2022005431 (ebook) |
ISBN 9780231195867 (hardback) | ISBN 9780231195874 (trade paperback) |
ISBN 9780231551526 (ebook)
Subjects: LCSH: Art—Psychology. | Aesthetics—Physiological aspects. |
Neurosciences and the arts.
Classification: LCC N71 .C28513 2022 (print) | LCC N71 (ebook) |
DDC 701/.15—dc23/eng/20220607
LC record available at https://lccn.loc.gov/2022005430
LC ebook record available at https://lccn.loc.gov/2022005431

Columbia University Press books are printed on permanent and durable acid-free paper.
Printed and bound by CPI Group (UK) Ltd, Croydon, CR0 4YY

Cover design: Milenda Nan Ok Lee
Cover art: © Walid Raad

# CONTENTS

# ACKNOWLEDGMENTS

This book has benefited enormously from exchanges with colleagues and friends over the ten-plus years since the publication of *Neuroestetica: L'arte del cervello*. Without some of them, such a new and expanded opus would never have seen the light of day. Fernando Vidal was the most discerning reader and most gracious interlocutor I could ask for. With him I organized the international conference "Le sensible à l'oeuvre" at the Centre de recherches sur les arts et le langage (CRAL) at the Ecole des Hautes Etudes en Sciences Sociales (EHESS) in Paris, during which time I cleared up many of the interdisciplinary challenges of the inquiry. Esteban Buch is to be thanked for supporting the initiative. In these efforts I was also aided by the Fondation Calouste Gulbenkian, headed by Joaõ Caraça, which hosted my exploratory workshop "Arts, neurosciences et sciences cognitives." I am deeply grateful to David Freedberg and my fellowship at the Italian Academy for Advanced Studies at Columbia University in New York for several rich and wide-ranging discussions on the make-believe and embodied simulation, which kept me from falling into an intellectual comfort zone. The open mind with which Gretty Mirdal invited me for a stay at the Paris Institute for Advanced Study to explore the relationship between art and neuroscience, and the reception she gave me, are memories I cherish with gratitude and hope to repay by repeating her gesture for the next generation of women scholars. Annalisa Sacchi and I wrote an entire project titled

"Aesthetics and Neurosciences: The Neurophysiological Correlates of Fiction," with the thoughtful support of Alberto Priori from the neuroscience side. At the time, few were willing to credit the topic; she and I have since learned a great deal from each other.

Stefanos Geroulanos provided a key spark. He was persuaded that this work would have a North American readership; I hope I have proved him right. Andrea Pinotti and Graham Burnett have shared many insights and supplied constant intellectual stimulation over the past decade. Barbara Grespi's critical and enthusiastic reading of some passages that are particularly important to me in chapter 5 was highly valued. Barbara Carnevali gave me thoughtful tips on matters pertaining to the general setup of the book. Giandomenico Iannetti's precise arguments convinced me I was justified in regarding neuroscientific texts as elaborate pieces of rhetoric. Gabriele Sofia offered ready and generous advice during the lengthy process of bibliographic research. Andrea Soffientino retrieved essays and articles for me from a vast range of online databases, and it was wise that I never asked how he did it. When the research was nearly finished, Jessica Murano opened my eyes to Angelo Mosso, a compelling author I could not pass up including. Nicole Miglio was a faithful and exacting reader, able to detect glitches in an argument and tell me in a way that helped me rethink it. I have Marta Calbi to thank for all the mistakes I avoided while dealing with brain mechanisms and neuroimaging. Many ideas would never have made it to the page were it not for my conversations with these two, along with the other members of my Performing Identities Studies (PIS) research group, notably Giancarlo Grossi, Irene Pipicelli, Sofia Pirandello, Giulia Rignano, and Samuele Sartori. Silvia Romani is both a dear friend and the best coach I could hope for.

To Wendy Lochner, who welcomed this text into the series she edits, goes my sincerest gratitude for the sensitivity, patience, helpfulness, and confidence she showed me during the final writing phase, which coincided with the pandemic. Without Samuel Fleck, this volume would never have been written. Not only was Samuel a precise and rigorous translator of Italian and French, able to move elegantly between registers and between a scientific and a philosophical lexicon, but it is to him I owe the clarity of many passages, the careful editing, the direct verification of sources in English, the meticulously assembled notes and bibliography, and the vision of the whole. That is not to mention the invaluable professionalism and courtesy that accompanied his work over the past year and a half, during which he proved an ideal sparring partner. Walid Raad has been friendly and generous in making available his *Preface to the third edition (édition française) _ Plate III* (2016) as a compelling cover to introduce my book to its readers, and Luyang's *Delusional Mandala* in the frontispiece has helped me present today's "brain anxiety" in a not-so-somber light.

Without Olivier and the lovely support he demonstrated, even when I did not know I needed it, this endeavor would have been a far less joyful one.

This research was funded by the Department of Philosophy "Piero Martinetti" of the University of Milan under the project "Departments of Excellence 2018–2022" awarded by the Ministry of Education, University and Research (MIUR). The preliminary translation of the volume *Neuroestetica*, which constituted the basis for this new book, was co-funded by the Paris Institute for Advanced Study. Sources have been updated, arguments expanded, and all chapters substantially rewritten. The critical bibliography and the appendix of artworks have been compiled for the present version.

Milan, March 1, 2022

I believe, as St. Thomas did, that the brain is at the center of the world, that it is the cornerstone of the world.

Io credo, come San Tommaso, che il cervello sia al centro del mondo, sia il perno del mondo.

—Felice Casorati, *Un sogno*, conference at the Vieusseux Institute, Florence, 1953

"I'm trying to complicate the locatability of human identity as a here and now, an enclosed and finished product, a causal force upon Nature. Or even . . . as something within Nature. I don't want the human to be in Nature, as if Nature is a container. Identity is inherently unstable, differentiated, dispersed, and yet strangely coherent."

—Vicki Kirby, 2002, personal communication in Karen Barad, *Posthumanist Performativity*, 2003

# *NEUROAESTHETICS* RELOADED

*Ducunt volentem fata, nolentem trahunt.*

—SENECA, *LETTERS TO LUCILIUS*, 107, 11, 5

My previous work on neuroaesthetics, published in 2009, coincided with the Copenhagen Neuroaesthetics Conference. Summarizing that event, Marcos Nadal and Marcus Pearce described how they used "the term neuroaesthetics to encompass the study of the neural and evolutionary basis of the cognitive and affective processes engaged when an individual takes an aesthetic or artistic approach towards a (Western or non-Western) work of art (used in the broad sense to include music, film, theater, poetry, literature, architecture and so on), a non-artistic object or a natural phenomenon."[1] In those days, neuroaesthetics was clearly an umbrella term, an all-inclusive disciplinary label. Since then, research paths have specialized further, but we have yet to see anything like a final consensus on its epistemic status. At the same time, as a collective enterprise run largely by neuroscience research teams, it proceeds according to an incremental notion of knowledge production.

As I see it, neuroaesthetics studies the human being, scrutinized in vivo while thinking and feeling, through brain-imaging techniques. It therefore competes with more theoretical discourses about our cognitive and emotional processes and personal condition. Highly disconcerting for scholars in the humanities, this challenge has become a point of contention between philosophers, on the one hand, and those working in quantitative and experimental fields, on the other. Yet, after more than a decade, the neuroaesthetic enterprise is better poised than many other well-reasoned proposals to illuminate the current philosophical landscape. Indeed, it offers a litmus test for our contemporary understanding of media and (living) matter. How can we define our entanglement with the natural and artifactual environments? Through its stances on human life, image making, technical and artistic processes, and the role of the experimental setting, neuroaesthetics serves to tackle this question.

Hegel taught us that philosophy comes late, like the owl of Minerva. Is that still the case? Is neuroaesthetics the latest successor to Western philosophy, repackaging long-standing ideas about the Self in more recent ones about cyborgs and in technologically updated claims? Or is it actually involved in forging a new narrative of identity? These are the issues I raise in this book.

Of course, a philosopher can always be told: "You don't know the brain." In a manner of speaking, that is correct. I have never studied human neural mechanisms in all their enchanting detail. But I have encountered the brain in neuroscientific papers whose rhetoric is no less compelling than French theory's, and in images whose fictionality is no less effective than that of oil portraits. I therefore know the brain on which neuroaestheticians work and am addressing its agency as a cultural artifact.

Milan, April 4, 2021

# EMBODYING ART

# 1

# 1994

## Putting Neuroaesthetics on the Map

### CALL ME BRAIN

We live in a time of brain-based visual narratives. In 1983, *Vogue* published a series of position emission tomography (PET) scans of our cerebral matter; it was the first nonspecialized magazine to do so. The models had feathered hair, and the pictures supposedly identified normal, depressed, and schizophrenic persons.[1] The intent was to make readers aware of modern diagnostic devices. In 2003, conceptual artist Jonathon Keats underwent functional magnetic resonance imaging (fMRI) while thinking about art, beauty, love, and death. The resulting images, exhibited at the San Francisco Modernism Gallery, served a novel purpose: selling his brain, together with his original thoughts, which he marketed as an artwork created by his own cerebral activity.[2] A growing alliance is in place between image making and the scientific enterprise aimed at determining neural processes of subjectification: I am the one who looks at my brain. It would seem that neuroscience has saturated the explanatory field of the Self, aided by depictions of cortices, thalami, and the like.[3] In 2007, brain imaging had evolved to the point that the American Psychological Association saw fit to celebrate fMRI as a technique that "produces movies starring the brain."[4]

Rigorous scholars, journalists, and opinion makers are all engaged in a neuroculture that "encapsulates social and cultural values that arise and evolve with our understanding of the nervous system."[5] They participate in a common "brain talk,"[6] which has become so pervasive as to prompt calls to "de-neurologize."[7] While some neuroscientists consider the way "the media communicate, or even distort scientific discoveries [to be] totally beyond [their] control,"[8] others argue that "they have a responsibility to . . . offer antidotes to the press' tendencies to simplify, exaggerate, and dramatize findings."[9] In either case, as the historian of science Fernando Vidal makes clear, neuroscientists "seem to consider the sciences as *having* 'social implications' or an 'impact' on society, rather than as *being* themselves intrinsically social activities that prosper largely through strategies embedded in the social fabric,"[10] as they in fact are. From a social sciences perspective, Alain Ehrenberg contends that the neurosciences "belong to the general dynamic of treating the patient as an individual conceived of as the agent of his or her own change, as in [that] variant . . . of autonomy that is American empowerment,"[11] to the point where neurobiology becomes a matter of self-management.[12] The end result is an embrained "I."

We cannot, however, limit ourselves to evaluating research approaches based on their capacity to enchant or repel us, or wagging a finger at the often uncritical and sometimes sloppy application of neuroscience to everyday life and personal experience.[13] That sort of reaction would simply confirm the magic enacted by neuroscientific explanations, which prove effectively alluring for many reasons. As neurosurgeon Bree Chancellor and cognitive neuroscientist Anjan Chatterjee explain, "neuroscience language is reductionist and is appealing because it is concrete and appears technical and objective," and "brain imaging studies have disproportionate credibility in the scientific community as well as with the public."[14] It is to this disproportion that we ought to direct our attention. It rests on the mismatch between the living, organic brain, which is the object of neuroscientific study, and the brain whose adventures we follow in TV series such as *Perception* and *Black Mirror*, the one that we motivate and reeducate with Brain Bullet software,[15] that we enhance with video games like *Brain Training*, that has a health we pay companies to insure, and that neural capitalism exploits through the "new alliance between brain science and technology."[16] Indeed, even some of the most dedicated neuroscientists acknowledge, with some perplexity, that "brains make money."[17] It is also the brain that gets better by "viewing art, even if you know nothing about what you're looking at."[18] *This* brain is the result of digital renderings of graphic data that portray it; it is a rhetorical and manufactured artifact.[19] This culturally imbued matter is what neuroaesthetics addresses, what it enhances, and whose short history overlaps it.

The endeavor to make this most precious hidden organ visible dates to antiquity. It even predates the introduction of the term *neurologie*, which came into English in 1681, via Samuel Cordage, in his translation of Thomas Willis's *Cerebri anatome* (1664), the foundational text of neurology.[20] Five anatomical drawings, preserved in a manuscript from the year 1200 but backdated to 300 BCE and traceable to Alexandria, Egypt, depict the brain organ inside the skull, irrigated with blood. The visualization process was resumed only after the centuries-long hiatus brought on by the prohibition of human anatomical dissection. The oldest extant original drawing of this organ is from an eleventh-century manuscript belonging to Caius College, Cambridge.[21] The turning point, however, came later, between 1504 and 1507, at the hospital of Santa Maria Novella in Florence, when Leonardo da Vinci injected hot wax into a specimen of cerebral ventricles to make a cast of the brain and used this to create a detailed drawing of the gyri imprinted on it. What distinguished Leonardo's undertaking was the search for scientific evidence of natural findings, coupled with the implicit idea of human exceptionalism. Today neuroaesthetics makes us rethink our *unique naturalness*, objectifying our cognitive functions while at the same time investigating our involvement with what is most contingent and fabricated: artworks. I will present theoretical arguments and examples from the contemporary visual and performance arts to make the case that this incipient discourse, owing precisely to the tension between biological reductionism and cultural production, helps dismantle enduring albeit rearguard assumptions about the separateness of nature and culture, which extend to their respective fields of research.

Charles Snow, an undistinguished experimental scientist better known for writing novels, canonized this opposition with an entire thesis about the existence of two cultures. In a 1959 essay, he addressed the mutual distrust between fact-based scientists and imagination-driven humanists, remarking how English engineers considered attempts to read Dickens a more than sufficient oblation at the altar of so-called culture and, conversely, how any renowned humanist, if asked for the second law of thermodynamics, would respond by claiming ignorance. Snow's *Two Cultures* turned him into the spokesperson for a widely held consensus on the matter and is still cited today by those seeking to bridge this divide. Eric Kandel, Nobel laureate and prominent neuroaesthetician, mentions the book at the opening of his *Reductionism in Art and Brain Science: Bridging the Two Cultures* (2016), in which he promotes an understanding of both art making and experiment design. Developing research in neuroaesthetics, Kandel suggests, will allow us to overcome mutual disciplinary deafness.[22] As I have just suggested, the hypothesis has merit. Snow's reasoning was grounded in an anthropology of personal encounters and confrontations between different intellectual personalities, the traits of which are assessable via questionnaire, as validated by a study conducted several decades after his book was published.[23] Interestingly enough, scientific personalities are exactly what William Hirstein, a major neuroaesthetician whom I will discuss in the next chapter, considers the crux of the neurocultural enterprise: "The move to neuroscience . . . brings with it the idea that vast amounts of information are missing, and in the process of being revealed, and that the researcher of the mind must begin a long apprenticeship in neuroscience in order to begin to access the mountains of information it is now producing, something that most philosophers are uncomfortable doing."[24] Philosophers, it would seem, fall into the lazy camp. It is astonishing to find the very same allegation leveled from across the aisle at seminars and conferences and in personal conversations. The sense of research being about personality traits is only reinforced by conciliatory proposals that look ahead to "partnerships" between neuroscience and the humanities,[25] and calls for a "co-dependence thesis" between the two cultures (which neuroaesthetics especially would enable) given that "theorizing does not happen in a vacuum"[26] and researchers are living, breathing people. Epistemic challenges resemble personal disputes. This happens for two main reasons. First, scholars are still striving to define shared epistemic criteria. As James Croft has rightly observed, "researchers in the humanities simply misunderstand . . . and misuse . . . neuroscientific findings and scientists fail . . . to meet the disciplinary standards of the humanities."[27] Second, when scientific research goes public, which today occurs at a rapid pace, every researcher is expected to take the floor in the first person to disseminate findings and garner attention and approval. The fact that problems and methodologies overlap without merging has all but prevented "the faith of neuroscientism,"[28] which of course is a matter of personal preference. The same goes for neuroaesthetics as goes for the vast field of neurohumanities, to which it belongs.

Of course, personal controversies and clashing interests have always made up the mundane side of the disciplinary conflicts that accompany the emergence of a new field. This emergence can be acknowledged and discussed in two distinct ways. One is by conducting a bibliometric analysis of publications tagged by keywords to develop an "unbiased" account of the state of the art.[29] The other is by approaching the battlefield of live knowledge production. As Roland Barthes wrote back in 1971, "interdisciplinary activity, today so highly valued in research, cannot be achieved by the simple confrontation of specialized branches of knowledge; the interdisciplinary is not a comfortable affair: it begins effectively (and not by the simple utterance of a pious hope) when the solidarity of the old disciplines breaks down—perhaps even violently, through the shocks of fashion—to the advantage of a new object, a new language, neither of which is precisely this discomfort of classification which permits diagnosing a certain mutation."[30] This is my take on neuroaesthetics. Despite being granted an entry in the *Oxford Encyclopedia of Aesthetics* since 2014, neuroaesthetics is far from achieving canonical status. The struggle is not over. At the same time, it "holds to operational definitions,"[31] and its outputs are exploitable on a social level to facilitate funding for the costly laboratory experiments on which neuroscientific research is largely based and for which grants, public or private, must be obtained.[32] Moreover, it is fashionable, as its ramifications in neurourbanism[33] and environmental neuroaesthetics illustrate.[34] But to understand how this enterprise can be "poised to become a mainstream scientific topic of inquiry"[35] without yet being an established research field, we must turn from day-to-day research to epistemic questions, updated for the current theoretical and media landscape.

The disjunction between the humanities and the sciences is not so much about the role of personal imagination versus rigorous modeling, or about profitability. It has to do, more than anything, with the distinct way each domain is studied, received, and implemented. The objection raised by philosopher Giulio Preti in 1968 remains valid to this day: the problem stems from "two different scales of values, two different notions of truth, two different discursive structures."[36] This entails two forms of understanding, two alternative styles of writing, two ways of generalizing experience, if not two divergent areas of research freedom, each granting access to a seemingly heterogeneous branch of knowledge that responds to distinct conditions of thinkability. It is about disciplinary rhetoric and apparatuses. Art historian James Elkins calls it "the problem of the pencil":[37] whoever picks up a literary essay, an account of an archaeological excavation, or a work of art criticism can read straight through the text, yielding to the rhythm of the narrative; the reader of a scientific text, by contrast, has to work out the proposed experiments and formulas at first hand. At first glance, the two strategies seem to mirror the original polemical kernel that in various ways has set the rhythm for Western knowledge in its succession of questions, challenges, and schools. It began with Plato and Aristotle; Raphael's celebrated fresco *The School of Athens* shows the former with a raised finger pointing at the sky and at the immutable forms of knowledge and truth, while the latter stands beside him with a lowered arm and an open hand, held out to cover earth's extension, the bottom-up source of empirical knowledge. The contest has persisted through the ages, playing out under diverse guises. A twentieth-century variation of this dispute pitted the "rigorous" analytic philosophers, stewards of a disembodied and logical language, against the continental philosophers, "emotional" thinkers of the first-person experience. That contrast, in turn, echoes the seventeenth-century quarrel between the Moderns, progressive exponents of investigation and critique, and the Ancients, erudite and conservative scholars of the passions. Of course, the history of ideas is made up not just of pitched battles but also of karst landslides, as when hermeneutics prevailed over positivism in France, neopositivism defeated hermeneutics in the United States, and the United States became the birthplace of French theory. The whole economy of precursor and successor figures shows the rhythm of knowledge to be far more pendular than progressive.

In the context of neuroculture, this dialectic is instantiated by the controversial relation between the life sciences and the human sciences, notably philosophy, which permeated Western culture from the nineteenth century through the very recent past. Although it was already criticized in 1998 by neurologist Jean-Pierre Changeux in dialogue with philosopher Paul Ricoeur, prompted by their mutual conviction that "the institutional gap that separates the life sciences from the humanities and social sciences has had catastrophic results,"[38] and it is currently under attack by cultural anthropology, on the one hand,[39] and the 4E Cognition approach—which stipulates that "brains cannot be divorced from their bodily and environmental context"[40]—on the other, this divide still plays a decisive role in our common sense. The precise terms of the dichotomy entered the canon in 1883 thanks to Wilhelm Dilthey, who, in his counter-attack on positivism, wrote: "The independent position of such a [human sciences] discipline cannot be contested, so long as no one can claim to make Goethe's life more intelligible by deriving his passions, poetic productivity, and intellectual reflection from the structure of his brain or the properties of his body."[41] Along with Hegel, Goethe was the watershed in the crisis between science and philosophy, which thereafter, according to Ernst Cassirer, found clear expression in the rift between natural and cultural sciences. Such a split hinges on the facticity of the positive sciences, which Edmund Husserl retraces, in *The Crisis of European Sciences* (1936),[42] to the gesture whereby Galileo mathematized nature. Since that time, the scientific method has advanced by building models, some of them mutually exclusive, to be tested incrementally via verification processes. The Galilean approach inaugurated those mere "fact-minded sciences" that give rise to mere "fact-minded people." And yet Galileo as a figure epitomizes the extent to which such fact-minded people are subject to sense. How come Galileo knowingly disregarded Kepler's first law, which dictates that the planetary system is heliocentric, has elliptical (as opposed to circular) orbits, and takes the sun as one of two foci? As the art historian Erwin Panofsky explains in *Galileo as a Critic of the Arts* (1954),[43] this is because the physicist chose the Renaissance circle over the mannerist ellipsis, just as he preferred Ariosto over Tasso and favored clear images over anamorphosis. Moreover, unlike Kepler, who believed that the human body followed a principle of rectilinear movement, Galileo considered human movements to be circular. Galileo—a talented sketch artist—took the same view on the matter espoused by Leonardo in the *Treatise on Painting* and ignored the contributions of his fellow astronomer. Panofsky plainly shows that cultural preferences were as decisive for Kepler's acceptance of ellipses as they were for Galileo's rejection of them. To be sure, Galileo was more quantitative and less animistic than Kepler—and more modern, for that matter—but he was ultimately driven by stylistic preference. Any attempt to give a tidy and purely conceptual account of research dynamics ends up simplifying and forcing the issue. Was it not Freud, the father of psychoanalysis, a neurobiologist by training—credited with publishing studies on the nervous system and devising a histological method for staining its pathways—who drafted the first theory of the neuron?[44] According to Kandel, Freud even had a "lifelong identification with fundamental, positivist science."[45] Does Paul Feyerabend, a philosopher of science, not use art historian Alois Riegl's *Late Roman Art Industry* (1901)[46] to argue that science, no less than art, evolves according to styles?[47]

Ironic though it may seem, the project of reconciling biographies with research methods and intellectual environments, cognitive preferences with epistemologies and medial and material devices, strikes me as precisely the cultural scope of neuroaesthetics: rationality as such is emotional, and people's thoughts, desires, and performances are historically grounded and displayed in an artifactual, if not an artistic, context. According to neurobiologist Antonio Damasio, a *pure* reason, cut off from the materiality of the body and its natural and social bonds, free of contradictions, and impervious to emotional stimuli and circumstances, would produce a pathological intelligence and incoherent life habits. He writes in *Descartes' Error* (1994): "the cool strategy advocated by Kant, among others, has far more to do with the way patients with prefrontal damage go about deciding than with how normals usually operate."[48] Thus, the scientist maintains, we are required "to explore the threads that interconnect neurobiology to culture."[49] Such a neurocultural enterprise ought not to neglect the history of philosophy and would only be strengthened by the knowledge that Descartes conceived of the *res extensa* not as a personal substratum but as an epistemological notion and was, as his correspondence with Princess Elisabeth of Bohemia unmistakably shows, perfectly aware of the psychosomatic nature of his fellow human beings. No less aware was Dilthey: "the mental life of a man is part of a psychophysical life-unit which is the form in which human existence and human life are manifested. [But] only by means of abstraction is mental life separable from that psychophysical life-unit."[50] I applaud the neurocultural effort, insofar as it is legitimate to simultaneously be opposed to its reductionism *and* espouse a materialist approach whereby brains and bodies "are not essential, fixed or universal" but, rather, "[are] shaped in and through experience with other body-minds, objects, and entities in the world."[51] What matters is how exactly this epistemic enterprise is put into play, according to what discourse, under what knowledge regime, and with what kinds of artifacts and apparatuses.

Neuroscience is playing a part in what has been called the "third culture,"[52] in which working scientists write directly for laypeople, and—on a higher level—is enhancing the so-called biocultural turn. It deals with the interplay of natural processes, cultural performance, living beings, and artifacts, including robots and AI devices, by targeting the shifting ground of what it means, or feels like, to be living and situated human beings. The biocultural paradigm conceives of all human technology as simultaneously expressing the cognitive capacity and the bodily nature of our species. As Cliodhna O'Connor and Helene Joffe point out, such a turn should not be considered groundbreaking per se: "New scientific information can indeed challenge and modulate existing understandings; however, it can also assimilate into and reinforce established ideas. It is therefore not self-evident that neuroscience will substantively alter understandings of personhood in predictable directions."[53] After all, brainocentrism's consistency with previous philosophical paradigms is backed by evidence. "The cerebral subject," as Francisco Ortega and Fernando Vidal have called it, predates reliable neuroscientific discoveries. It first appears in modern Western philosophy in John Locke's 1690 *Essay Concerning Human Understanding* and "has all the appearance of having been a motivating factor of brain research"; "as it advanced, this research legitimized and reinforced the brainhood ideology" but did not trigger it.[54] Brainocentrism is in no way science based. According to Vidal, "the idea that 'we are our brains' is not a corollary of neuroscientific advances, but a prerequisite of neuroscientific investigation."[55] Some are skeptical about the pervasive degree to which the brain has become synonymous with the Self and argue that, in fact, under no circumstances do "individuals consider themselves to be mere fleshy puppets of their brains,"[56] even in the context of psychopharmaceuticals, whose coarse mode of action, incidentally, suggests that we know far less about cerebral mechanisms than was previously thought. Taking brainhood as an emergent facet of selfhood could even be considered a democratic gesture, given that everyone has access to his or her brain, while many lack access to education. But I do not think the most appropriate epistemic move is to slow down and take it easy, as if we must take care of our brain the way we do our liver. Nor should our tendency to appeal to different ontologies concurrently in our lifetime distract from the fact that we privilege our brain as an agent endowed with the specific capacity to make us feel, think, and act. Thoughts, emotions, and actions are understood as embrained, and this is so despite clear biological evidence to the contrary.[57] O'Connor and Joffe are then justified in claiming that, "given the significance of folk psychological understandings in guiding everyday behavior, perception and social interaction, examining neuroscience's influence on commonsense conceptions of personhood is arguably a more pressing task than establishing whether public understandings of the brain are scientifically correct."[58]

Is Catherine Malabou mistaken in asserting that "the brain has never been an object of philosophy"?[59] Her argument holds true for the past, when the brain was considered the material instantiation of the Self but was not regarded as an autonomous subject, an agent, as long as it was an intracranial matter, or in Locke's words, "the mind's presence-room."[60] With the advent of brain imaging, we stopped needing to trephine the skull to gain access to it, and it became a matter for visual culture as well as for neuroscience. Our cerebral organ—our "inner space," as Kandel plainly calls it[61]—stepped out of the skull to appear publicly as the protagonist in a new narrative of the Self. Neuroaesthetics offers a fruitful perspective from which to understand our manufactured, visualized ego.[62]

## BETWEEN NATURALIZATION AND ARTIFACTUALITY

The mainstream consensus is that neuroaesthetics originated in 1999 with the seminal work by neurologist Semir Zeki, *The Inner Vision: An Exploration of Art and the Brain*, although we also find competing claims. According to Changeux, the term first arose at the 2002 conference "The Pleasure of Art as Sensed by the Brain," held in San Francisco.[63] More recently, Camilo J. Cela-Conde and Francisco J. Ayala, two committed neuroaestheticians, have contended that "beyond some valuable precedents [such as Ramachandran and Zeki] the empirical field of neuroaesthetics started in 2004,"[64] thanks to studies by Oshin Vartanian and Vinod Goel, Hideaki Kawabata and Zeki, and Cela-Conde and colleagues.[65] I would argue that, in fact, neuroaesthetics emerged in the early nineties as a collective, albeit not a coordinated, enterprise, and the variety that characterized its inquiries at the beginning remains to this day. As psychologist Claus-Christian Carbon bluntly put it just a few years ago, "current research attempts are mostly unconnected to each other, even within one research group or even across different studies of one single researcher—they mostly lack ideas to connect different results and to comprise them by a more general theory on aesthetics."[66] In 1994 and 1995, three texts on the visual arts were published by scholars familiar more with brain studies than with art theory: *The Neurology of Kinetic Art* by Zeki (with visual artist Matthew Lamb) in England; *Raison et Plaisir* (Mind and pleasure) by Changeux in France; and *Arte e cervello* (Art and brain) by Lamberto Maffei, working alongside Adriana Fiorentini, in Italy. The authors—despite knowing one another from past research—had been pursuing their endeavors independently. To this constellation may be added the aforementioned book by Damasio, who also wrote short essays on the visual arts;[67] as director of the Brain and Creativity Institute at the University of Southern California, he acknowledged that his dream (shared by his wife, Hanna, also a neuroscientist) was "to find a way to apply the new science of the mind emerging from neurobiology to art,"[68] and his research has indeed been pursued along those lines.[69] Kandel would agree with him: "One of the aspirations of this new science [neuroaesthetics] is to link the biology of the brain to the humanities. One of its goals is to understand how the brain responds to works of art."[70]

Those were thus the years that put neuroaesthetics on the map. Of course, this enterprise is not entirely without precedent. Science and art have served each other frequently and regularly, as masterfully exemplified by the seventeenth-century French painter Charles Le Brun's sketches of the face, which were used for studies on human physiognomy, and the nineteenth-century chemist Michel Eugène Chevreul's theory of colors, which was applied to Georges Seurat's *La Grande Jatte* (1884–1886). What is new, however, is the high degree to which neuroscientists claim that they can *explain* what artworks do to and for our cognitive processes. For the initial players involved, neuroaesthetics marked the dawn of a neurology of aesthetics, whereby it became possible to understand the biological bases of visual—and possibly artistic—artifacts. It ushered in what John Hyman straightforwardly describes as "the study of art by neuroscientists":[71] to speak of neuroaesthetics is to speak of *visual* neuroaesthetics. In 1999, Zeki was convinced that he found himself "at the threshold of a great enterprise,"[72] and his pitch was little changed some twenty years later, when he argued that "neuroaesthetics seeks inspiration and insight from works of art and from debates in the humanities to try to gain some insights, however small, into the workings of the brain."[73] Maffei, who, like Zeki, spoke of a "visual brain," took a similar view, writing that neurophysiology and neuropsychology "do not offer solutions, but lay the foundations for speculations, . . . to bridge the gap between scientific knowledge and the visual arts."[74] It was the beginning of an effort that also saw important contributions from Vilayanur S. Ramachandran, who in 1999 proposed a "neurological theory of aesthetic experience"[75] and in 2003 announced the release of *The Artful Brain* in the published conference proceedings for a series of Reith Lectures.[76] The pioneering spirit of these early investigations was shared by Changeux, who called on others "to pursue the work initiated by Gombrich [, which] can be achieved by investigating the possible neural origin of aesthetic pleasure and artistic creation, and jointly by reevaluating the evolution of a given painting."[77]

The art historian Ernst Gombrich is, along with Rudolf Arnheim, an important reference point for the field. This appeal, however, stems from a gross misunderstanding of Gombrich's search for stable patterns in art production and spectatorship as requiring or supporting a search for preconscious cerebral mechanisms.[78] To assume as much would be to disregard Gombrich's argument that image making comes before matching—that is, that artists make artworks in an experimental and even improvisational way, confront the results with what they see, and then adjust them to it,[79] and correspondingly, that "beholders make their own sense of pictures by collating what they see on the canvas with what they know about the world and with what they remember of other pictures."[80] Neuroaesthetics's conviction that artistic production and aesthetic beholding are purely neurological initially led it to claim an anti-technological bias, although this quickly came into conflict with certain major findings about the brain, such as brain plasticity and the dependence of epigenesis on interactions with the environment; it also impeded discussion of how the artworld had itself appropriated brainhood as a poietic and stylistic element. The notion of a "cerebral art" had already been dreamed up in 1914 by the first film theorist, Ricciotto Canudo. His *Manifeste de l'art cérébriste* (Manifesto of cerebrist art), published in the French newspaper *Le Figaro*, was enlivened by the idea that "art has been gradually and intensely cerebralizing for some thirty years or so." According to Canudo, "Baudelaire was the forerunner of this subtly cerebral aesthetic that found its two best expressions in the precursors of our lyricism of today and tomorrow: Rimbaud and Mallarmé [who] first asked prosody for a new emotion, one capable of making the brain quiver more than the heart."[81] The same would apply, in his view, to sculpture and the visual arts. Gilles Deleuze draws on brain mechanisms in *Cinema 2: The Time-Image* (1985), calling cinema "the cerebral art par excellence, the internal monologue of the brain-world."[82] The neuroaesthetic assumption that we must relate to the brain as an unmoved mover of human processes is nevertheless perfectly consistent with the field's beginnings as the expansion of neuroscience into a domain as yet unknown to it.

Owing to their view that art should primarily be explained neuropsychologically, neuroaestheticians quote Gustav Theodor Fechner, via his 1871 *Zur experimentalen Ästhetik* (On experimental aesthetics), together with his twentieth-century American counterpart, Daniel E. Berlyne, as one of their few regular scholarly sources. Such a conviction finds its precise counterpart in the idea that art is what is established as such. In both cases, art has a conventional definition. Neuroaestheticians' tendency, when dealing with the humanities, to appeal to contemporary art theorist Arthur Danto as a principal reference, given his institutional theory of art, is consistent with this.[83] The use neuroscientists make of Danto is highly instrumental, completely neglecting his clear stance that "history supervenes on perception,"[84] but it accords with their primary approach to art. Stating that the "artistic" quality of something depends on a social decision grants neuroscience free rein to discover what art really is. This, of course, is a vicious circle. Therefore, while philosopher Alva Noë discredits whatever neuroaesthetics does on the grounds that "an account of how the brain constrains our ability to perceive has no greater claim to being an account of our ability to *perceive art* than it has to being an account of how we perceive sports,"[85] many neuroscientists deem this a major asset to their research. They do so to the point that aesthetics *tout court* becomes "disentangled from research on art" and focused instead on hedonic appreciation and neuroeconomics. Such is the case because "empirical evidence shows . . . that the neurobiological processes engaged when people form pleasure and displeasure responses for works of art overlap with those engaged by nonart stimuli."[86] The methodological problem stands: is it necessary to establish ahead of time what art is? We will see how this apparently intractable problem actually offers a special analytic space for interrogating the fictional quality of the experimental setting.

Let us return, however, to discussing the lineage of the incipient field. Neuroaestheticians, caught between an ambition to be groundbreaking and a desire to claim lofty origins, sought to position themselves in the wake of *Gestaltpsychologie*, which developed between the end of the nineteenth and the beginning of the twentieth century and exerted influence on the fields of art history and aesthetics, particularly the visual arts. It is especially unfortunate that the works of German psychological aesthetics of the late nineteenth and early twentieth centuries remain largely untranslated and little known outside the world of specialists, since their research would greatly benefit the project. Particularly fruitful would be the works of Robert Vischer and his father, Friedrich Theodor Vischer, Johannes Volkelt, and Theodor Lipps, the last of whom in 1906 referred to aesthetics as a "field of applied psychology."[87] An indispensable reference for those theorists was the psychophysiology of Hermann von Helmholtz (himself considered a founding father of today's neuroscience[88]), who in 1892 gave a speech to the general assembly of the Goethe Society in Weimar in which he stressed Goethe's poetic and knowledge contributions, promoting a similarity between artistic and scientific intuition. August Schmarsow spoke to the general assumption of this constellation of thinkers: "Any artistic configuration that man could possibly attempt requires a sensible substratum deriving from his possession of the common genetic material inherited from mother Nature. This is indeed *aisthesis* [sensation], that receptivity of our nervous system, held in such high regard by the Greek thinkers, vis-à-vis stimuli to which we owe all our sensations and therefore our lived experiences."[89] French aesthetician Marc Jimenez is therefore absolutely justified in writing that "neuroaesthetics could be defined as 'neuro-*aisthesis*,' focused on the sensitive body."[90] More challenging is the other way around, when neuroaestheticians, aiming to assimilate their positions to the principal notions of Western aesthetics, try to reduce various specific approaches to a common denominator. This critical effort mirrors, on a heuristic level, the idea that the brain is the unique and effective agent in human history and that mastering it is essential for any attempt to explain human behavior. In 2014, Marcos Nadal, Antoni Gomila, and Alejandro Gàlvez-Pol collaborated on an ambitious essay, "A History for Neuroaesthetics," in which they suggested that neuroaesthetics develops eighteenth- and nineteenth-century continental aesthetics, blending the experimental aesthetics of Fechner and Gestalt psychology with Kant's subjective judgment of taste and Darwinian evolutionary aesthetics.[91] This proposal, quite daring on its face, reveals the greatest epistemological problem at stake in neuroaesthetics. While it is true that the field has adopted an "empirical, quantitative, and reductionist character,"[92] equally true is that it is still seeking a research strategy epistemically richer than the statistical translation of subjective experiences—such as the fact that knowing the title of a painting changes our understanding of it—with whose significance and recurrence we are already well acquainted.[93] As for the community of scholars coming from the humanities, they remain wary of a horizon of thought and research that doubtless portends a great deal of ambiguity and more than a few approximations, owing primarily to the risk of crude reductionism. As a rule, we distrust reports of experiments establishing that the right hemisphere is less skilled than the left at recognizing and storing surrealist paintings[94] or that right-handed people have an "aesthetic preference" for images exhibiting a movement from left to right.[95] After all, what does our cerebral matter know about the anticipation that consumes us while we wait in line at the Louvre; or about the haste and resolve with which we proceed toward the famous hall, unheeding the scores of marvels on display in the rapidly traversed hallways; or, finally, about the modesty that, on our arrival, nearly prevents us from raising our eyes to the *Venus de Milo*? Maffei and Fiorentini argue that the attempt to explain aesthetic experience in neurophysiological and neurochemical terms is not a reductionist hypothesis because "all events of the emotional sphere and of consciousness can be traced back, in the last analysis, to the nervous system";[96] in other words, there is nothing beyond the nervous system, and therefore nothing to be added. What better case for reductionism could be made?

At the same time, however, much of aesthetics has thematized all along what is now called the embodied mind, performing in an environment inhabited by living beings and embedded with objects, as the plexus of those sensible possibilities that each human is. Science and aesthetics diverge because science deals with what is and what happens, and aesthetics deals with what *can* be and what *can* happen—specifically, when it questions the performative processes whereby the world engages us as humans in a reality that is held, known, and suffered in common. But ridding philosophical discourse of the level of reductive physical explanation does not entail having a subjective and relativistic understanding of cognitive processes, as long as the knowing of data is compatible with a plastic conception of knowledge, which, indeed, is enriched thereby. I therefore deem it wise to note that, notwithstanding the fancied distinction between the two cultures, many of the major aesthetically focused philosophers of the twentieth century relied on the knowledge science provided about the mind and body, primarily because, as Cassirer wrote in 1942, "a cultural object continually demands a physical-material substratum."[97] Friedrich Nietzsche, in critiquing Wagner's music in *The Gay Science* (1882), writes of "physiological objections" concerning his foot's freedom to skip, his heart's freedom to beat, and his blood's freedom to flow, all of which affect his bodily vitality; he continued developing a physiological aesthetics up through his last posthumous fragments.[98] Henri Bergson devoted the second chapter of *Matter and Memory* (1896) to a discussion of texts by neurologists of his time, showing a special interest in Wernicke's aphasia. Walter Benjamin based a considerable portion of his studies, notably *The Work of Art in the Age of Mechanical Reproduction* (1935), on the very idea that human perception and historical production coevolve.[99] Maurice Merleau-Ponty depended critically on contributions from Gestalt psychology in his *Phenomenology of Perception* (1945). References of this sort are no less abundant in the early classics of philosophical aesthetics. In *Critical Reflections on Poetry and Painting* (1719), Jean-Baptiste Du Bos expounds a climatic and physiological theory of genius, based on the idea that "the character of our minds and inclinations depends very much on the quality of our blood" and "the quality of our blood depends vastly on the air we breathe."[100] A similar focus can be found in Hippolyte Taine, whose *Philosophy of Art* (1865) describes the "essential character" of Holland—which Dutch art supposedly displays—based on its native climate, soil, and flora and fauna. Edmund Burke engages with the physiological knowledge of his day in the fourth part of the *Philosophical Enquiry Into the Origin of Our Ideas of the Sublime and Beautiful* (1757), where he investigates the "efficient causes" of the beautiful and the sublime. It is worth noting, in this connection, the principle of the succession of elements of uniform disposition, shape, and color, as exemplified in a series of uniform pillars, where, Burke claims, the eye is sent increasingly strong "impulses" with the observation of each successive pillar.[101] In 1774, Diderot even set to work on a text called *Éléments de physiologie* (Elements of physiology).

Finally, a more critical approach to the neuroaesthetic endeavor can help in understanding a discipline in search of recognition. An "emerging field of research"[102] in 2011, it was described in 2014 as still in its "early days"[103] and in 2016 as a "relatively recent field of research"[104] at a "historical inflection point."[105] Such hesitancy regarding its achievements is consistent with the diversity of approaches involved. Retrospectively, we can distinguish between an "experimental neuroaesthetics [that] produces data that are quantitative and vetted statistically" and a "descriptive neuroaesthetics [that] relies on observations that relate facts of the brain to aesthetic experiences. The claims are typically qualitative."[106] We can also approach the field by focusing on the three general systems of neural processes it addresses: emotion-valuation, sensorimotor, and meaning-knowledge.[107] Moreover, one of the major challenges faced by neuroaesthetics is resolving the imbalance between the untestable and the anecdotal, on the one hand, and the standard neural mechanisms, on the other. Let us concede, however, that having "little consensus on the concepts, predictions, or methodologies required" is quite common in the early stages of a new discipline.[108] Let us also put aside the regret expressed in 2019 by Martin Skov and Oshin Vartanian, who acknowledged that neuroaesthetics "cannot be said to have produced a commonly accepted definition of what differentiates an art experience . . . from experiences that employ identical computational machinery."[109] We should focus, rather, on three major points of agreement among neuroaestheticians. First, noninvasive neuroimaging techniques permitting empirical studies in controlled situations are the key element that made neuroaesthetics conceivable in the first place. Second, neuroaesthetics is grounded in "testable hypotheses,"[110] which are designed based mostly on visual artifacts. Third, as Helmut Leder and Pablo P. L. Tinio have emphasized, "not all objects are equally suited for studies using experimental methods": "In general, the better the relevant stimulus dimensions are controlled, the clearer the conclusion could be drawn from the data. . . . Thus, studies using real artworks may have to focus on finding highly salient effects that are not easily concealed or overridden by other variables."[111] Artistic agency is therefore at stake. My point is that neuroaesthetics reinforces the negotiation between naturalization and artifactuality. When experimental or theoretical contradictions occur, they shed light on the degree of fabrication proper to human life. I will therefore discuss the neuroaesthetic enterprise on a theoretical, historical, and critical level, by way of the material on which it focuses: images, bodies, and artifacts.

# 2

# NEUROAESTHETICS

## Cerebral Attributes and Bodily Ghosts

The search for principles and regularities that explain production and beholding in the visual arts breaks down into two main research strategies. The first investigates the conditions under which we see and the criteria that order the visual world—taken as a self-sufficient phenomenon for which paintings, scenery, and pictures in general primarily constitute *optical facts*. The second addresses the negotiation between the observer and the observed, which grants art a privileged position: it makes us see more, and in seeing more, see differently. Both are instantiations of "aesthetic experimentalism," through which "we can learn about the operations of perceptual systems by examining the productive strategies of artists."[1] Both take for granted that visual forms are representational or have the potential to be so because "they can be used in a narrative fashion to convey information about objects, people, and events."[2] While not always stated so obviously, this is the main cultural assumption at work in visual neuroaesthetic research. (This is not to say that neuroscience has ignored other art forms; on the contrary, it has dealt with music at length—to the extent that the neuroscience of music is the most developed kind of neuroaesthetic research today—although in doing so, it has "almost religiously avoid[ed] the terms 'aesthetics' and 'art' in its publication[s],"[3] thus eschewing the related epistemic issues.)

This assumption leads to two main, mutually contradictory stances. On the one hand, visual neuroaesthetics is grounded in "the parallelism approach": art mirrors the real world by mirroring the neural mechanisms through which we see and constitute it. By its own account, it "recognizes that the production and perception of art ought to conform to principles of neural organization."[4] It is based on this prior ordering that incongruences and discrepancies, despite occurring on multiple levels, are deemed to be such. The problem posed by such a *mirroring loop*, as I term it, is that images are supposed to refer to the "external world" but are in fact treated as equivalent to it. The seen world, and notably the manufactured one, is assimilated into its own optical structure. Visual neuroaesthetics, then, revives the prejudice of the "natural attitude," whereby, as art historian Norman Bryson writes, "the image is thought of as self-effacing in the representation or reduplication of things."[5] On the other hand, on an experimental level, the idea crops up, largely unthematized, that opticality and visuality do not simply coincide. Indeed, various models of neuroaesthetics test the subject's aesthetic behavior by selecting between naive and experienced groups of people, and they thereby acknowledge the image's status as more than just an embrained optical structure. The image here maintains a degree of cultural, material, and medial autonomy, in accordance with the knowledge we have of it and its making. Images occur in the world we inhabit together, not inside our heads.

Unfortunately, this contradiction remains generally unproductive because, as Chatterjee points out, "neuroscientists are likely to avoid definitional issues and focus on accepted examples of artworks or properties of these works as probes for experiments."[6] Against this background, it is instructive to see how different the methodologies and results can be. Chatterjee and colleagues, seeking an adequate tool for quantifying features in artworks, developed the Assessment of Art Attributes (AAA). Their conclusions suggest that "one need not be greatly concerned about enrolling experienced rather than naïve raters to assess descriptive attributes [given] the high inter-rater agreement in assessing art attributes."[7] By contrast, a few years later, Jane Else and colleagues argued that art expertise plays a decisive role in the response to visual artifacts.[8]

## THE EYE OF THE BRAIN

The first line of inquiry is that developed by Semir Zeki. In an essay published in *The Visual Neurosciences*, a major reference text for the neuroscience of vision, the British neurologist singles out color vision as an interesting research field on the grounds that "understanding how the brain constructs colors promises to give significant insights into understanding the cerebral processes underlying aesthetics in the old Greek sense, that is to say the acquisition of knowledge through the senses."[9] The term *aesthetics*—*aesthetica*, from the Greek *aisthesis*, glossed as "sensation" in the previous chapter—was coined in 1735 by philosopher Alexander Baumgarten, who, fifteen years later, referred to it as "the science of sensible knowledge"[10] and freed it from its state of infancy, confined as it was to expressing the volatility of the fleeting and unstable sense impression in individuals. The similarity between the two perspectives is no coincidence. No doubt Zeki's project shares the Baumgartenian spirit, which accords a cognitive value to our sensorium: sensations create an opening onto the world, and this may serve as a means of achieving intellectual knowledge, though it also bears knowledge in its own right. But Zekian neuroaesthetics does not constitute the final stage of German aesthesiology or English associationism; it is neither a "maximalist"[11] version of these movements nor the full flowering of Arnheim's lesson that perceptual and pictorial forms are the "the very flesh and blood of thinking itself";[12] nor yet is it an updating via brain imaging techniques of the ancient maxim *nihil est in intellectu quod prius non fuerit in sensu*. Rather, I regard it as a neurobiologically based theoretical project on human responses to stimuli from the outside world, approached through the medium of artworks, whereby, as philosopher Mark Rollins observes, art itself "*is* a science; a science of the senses."[13] In short, art equals aesthetic experience.

This line of investigation wields considerable influence over the neuroaesthetic enterprise as a whole, especially when it suggests that pictures depend more on "unconscious introspection" among artists and beholders than on the observation and study of the numerous depictions made available by art production. The "aesthetic" here becomes a feature of our brain. Luigi Francesco Agnati and colleagues indeed recommend that we think of it as "an attribute perceived by means of a particular brain processing system, in which the prefrontal cortex seems to play a key role [and whereby] human beings seem to share a common denominator underlying all artistic experience," regardless of any potential training in the arts. In other words, humans are thought to have "a basic knowledge of art,"[14] notwithstanding the fact that "humans" have never been questioned or examined on the matter transculturally, let alone diachronically, nor could they be, even if—as we will later see—neuroarthistory suggests otherwise. What characterizes the field is the assimilation of research question to experiment, concept to perceived object, artifact to stimulus, judgment to appreciation, artistic to aesthetic. The conditions of possibility of its inquiry are merely de facto operational options, with no thematized reference to the implied epistemic horizon. The result, I find, is that in neuroaesthetics hypothesis quickly becomes thesis. A study on the localization of the visual functions aptly illustrates this tendency: it treats Kandinsky as a noteworthy neurologist among painters, insofar as he "presents points, lines, or planes to patients, asking them whether they see the object presented."[15] As we read in *On the Spiritual in Art*, written between 1910 and 1912, the Russian master took part in experiments in the field of chromatic sensibility. In reporting the results of attempts to apply color to the treatment of nervous disorders, he notes that "red light stimulates and excites the heart, while blue light can cause temporary paralysis"; if this occurs for humans, animals, and plants, Kandinsky infers, it is because "colour can exercise enormous influence over the body as a physical organism."[16] Claims of this sort might seem to support the neuroaesthetic stance. Yet Kandinsky was writing in polemic against nineteenth-century materialism, with the intention of emancipating the sensible quality of the passions from the calculation of chromatic factors. This should serve as a reminder that artists' texts are not lab tests to be read in statistical-quantitative terms but working materials, poetic statements, and personal notes; they are artifacts in their own right. The same can be said for Leonardo da Vinci's observation—hardly the result of an anti-positivistic effort—that "the [first] principle of the science of painting is the point, the second is the line, the third is the surface, [and] the fourth is the body which is clothed by these surfaces."[17] However, against the background of such an understanding of human neural mechanisms, one pursuit mirrors every other: all cats are gray at night.

In such darkness, the notions of an "art-based neuroscience" and of the artist as a neuroscientist performing experiments in the studio instead of the lab lead to neuroaesthetic reductionism, which in turn triggers the naturalization of the image. Zeki does not discriminate among images, art objects, and tools. His assessment that Mondrian's compositions respond to the orientation-selective cells of the visual cortex, which "will be activated and will be responding vigorously" if the horizontal line falls to a cell responsible for horizontal lines,[18] applies equally to network patterns. By focusing on the pattern of the painting as an expression of the optical ability of the artist and the beholder, he abolishes all distinction, or collaboration, among form, content, style, and technique, in favor of optical agency as such. In positing visual artworks as neural expressions capable of shedding light on their own functioning as artistic stimuli, he altogether overlooks that artistic stimuli are handmade, not brain-made. Why, then, is the reference to artworks even methodologically necessary? Because these are conceived of in advance as extensions of brain functions, for which, to Zeki, artworks are mundane proxies. His reasoning is threefold. The first point concerns the role played by the unconscious, which he invokes without discussing. The second is the hypothesis "that physiological stimulation of specific visual areas can create aesthetic experiences";[19] although he concedes that we cannot thereby conclude that aesthetic experience depends entirely on brain activity, this nuance is eroded little by little in the course of his research. The third reason is that the image is completely emancipated from its outer reference: "in an important sense, a picture cannot represent an object; only the brain can do that, having viewed an object from many different angles and having categorized it as belonging to a particular class. A picture can merely imitate an object and then, . . . only one aspect of an object." Magritte, according to Zeki, proves as much in *The Use of the Word*, where a large, hyperrealist picture of a pipe appears above the sentence "This is not a pipe."[20] What Magritte portrays, however, is the exhaustion of the "as if" principle within figurative representation and the imminent failure of pictorial mimesis in art history; this is all the more obvious where the pipe *as a picture* does represent a pipe, just as in others of his paintings the apple depicts an apple, or the bowler hat a bowler hat, so much so that Max Ernst described Magritte's work as "the Reckless Sleeper method combined with photographic precision."[21]

But to acknowledge this pictorial project requires thematizing artists as something other than brains in a vat and fabricated pictures as something other than optical instantiations. It calls for an understanding of art production and spectatorship as performances and a notion of art "as a behavior ('artifying'), not as the result (paintings, carvings, dances, songs, or poems) or their putative defining qualities (beauty, harmony, complexity, skill)."[22] None of this is present in Zeki.

What I have called the mirroring loop overlooks the very principle of any aesthetic experience whereby the work disturbs me, moves me, upsets me, placates me, calms me down—whereby it interacts with me. Such a neglect is far more than a simple omission. It in fact perverts the very dynamics of art production and aesthetic beholding, which entail a permanent remaking and reframing. Writes art historian Whitney Davis, "when we look at things that have been actively configured for our seeing as actively configuring what is seen we aestheticize twice over or in a feedback loop, redoubling the aesthetic momentum of seeing: we paint the painting painted for our painting of it—repaint it."[23] But Zeki conceptualizes the experience of the imagined world from the viewpoint of a stand-alone brain. Even the retina obscures his argument, like a curtain over the sun. The simplicity and straightforwardness of his reasoning are what has made it exploitable, but, whether we accept it or not, this outcome ought not to be mandatory. Visual neuroscientist Patrick Cavanagh, for instance, takes a more cautious stance. He notes that the improbable physics found in certain paintings, particularly relating to the distribution of shadows, "suggests the physics of light and shadow used by our visual brain is simpler than true physics."[24] The fact that the pictorial process for creating an expressive effect in a painting is properly executed does not mean that the physical qualities on display replicate the physical reality scanned by our brain. For both Zeki and Cavanagh, however, not only are art production techniques *not* to be specially scrutinized, but artworks are not even specific artificial stimulants of the senses; rather, they are something that exists in a spatiotemporal vacuum. From this point of view, all that remains is to map art making onto neurobiological processes.

This is the position Zeki defends in his seminal article, "The Neurology of Kinetic Art." He establishes three laws. The first is that the image of the visual world is assembled in the visual cortex and not impressed upon the retina; the second is that the functional specialization of the visual cortex causes different features of the scene to be processed in separate regions of the cortex before being unified in a coherent image; and the third is that these elements ultimately consist of color, form, motion, and possibly depth. Such claims bear witness to the scientific battle he has been waging, since at least 1993 with *A Vision of the Brain*, against a conception of the visual system, long in vogue, that takes vision primarily to depend on the eye and secondarily to involve the entire brain or one of its parts—the visuomotor cortex. His perspective discards the earlier idea whereby, according to art critic John Ruskin, painting deals with the reflections of light and color on the retina. In its place Zeki advances a new description of visual functioning: area V1 receives stimuli from the retina and assigns the object a position in visual space; V1 is surrounded by V2, whose functioning is similar to that of V1; V1 and V2 are in turn surrounded by specialized areas V3 for direction and orientation, V4 for color, and V5 for the motion of the stimulus, in both the fronto-parallel and the egocentric planes and in both a centripetal and a centrifugal direction. All this is located in the posterior region of the brain. As PET scans show, the fact that one area is active in certain circumstances does not imply that the others are silent, although V5 is largely independent of V1. Vision, moreover, is articulated in successive phases: color is seen before form and form before movement, based on a modular perception[25] that unfolds at different speeds.[26] Such a distinction seems borne out by different kinds of blindness; for example, achromatopsia, or color blindness, is dependent on a lesion in V4, whereas akinetopsia, in which the patient cannot see objects in motion (and thus cannot pour coffee, being unable to tell when the cup is full, despite seeing the brown liquid "frozen in the air") indicates a lesion in V5. According to Zeki, there are also micro-consciousnesses, related to complete but partial perceptions. On this view, then, a conscious experience does not depend on the completeness of the perception: a patient who is blind as a result of a total lesion in area V1 would be able to consciously see fast-moving shapes thanks to an intact area V5. This is a kind of "blind vision." Leslie Ungerleider and Mortimer Mishkin of the National Institutes of Health account for it by distinguishing a "how" pathway, which is located in the parietal lobes and governs spatial functions such as taking objects whose position can thus be seen, from a "what" pathway, which is instead located in the temporal lobes and responsible for recognizing objects.[27]

Based on such knowledge, Zeki outlines what he terms his "credo," all the while disregarding the call by Changeux, from that same year, to consider neuroscientific findings from the time and treat any reflection on art creation and beholding as "a working hypothesis for current research, more than . . . an end point."[28] He detects a common pattern in image making across quite different artistic styles and brings together a broad range of pieces, from paintings and sculptures of Italian Futurism such as Boccioni's *The City Rises* to art objects such as Marcel Duchamp's *Bicycle Wheel*, Naum Gabo's *Kinetic Sculpture*, Jean Tinguely's irregular and comical *Metamechanics*, and Calder's mobiles. He takes up the heterogeneous group of works, whose creators all drew on the theme of motion, and redefines it in neuronal terms: if the visual brain processes stimuli in a modular way, then "aesthetics itself is modular,"[29] as artists (supposedly) know. This, Zeki maintains, is why the art for V5 does not consider form and color: a subject with a lesion in V5 would be unable to appreciate the fluttering of a Calder mobile (if it were moved by the wind) but could appreciate its redness. If V4 is the beholder of Fauvist art, as Susan Broadhurst claims,[30] then the V5 system is the true beholder of kinetic art. It is on these grounds that Zeki lays the foundations for a model of art theory: every picture is an image seen by the brain, and artworks function as super-efficient stimuli for inquiries into what I call *neuro-opticality*. Its (pseudo-)explanatory range is almost unlimited. Let us consider a painter whose work apparently resists interpretation as a static image, whose color backgrounds take on a spatializing function and figures defy analytic treatment—where, in other words, we find not lines but vectors of force: Francis Bacon. Deleuze has described his paintings as "hysterical."[31] This very same quality, the inversion and the distortion of the faces and bodies, prompts Zeki "to enquire whether cells representing faces and bodies on the one hand and objects on the other, are regulated differently, even if they co-occur in the same area(s) and whether it is because of this differential susceptibility that Bacon concentrated on deforming faces and bodies and sparing objects."[32] The implication is that painters, without exception, unconsciously use their expertise to achieve aesthetic effects by selectively stimulating visual areas of the cerebral cortex and thus provide material for future experiments.

This program suggests an intracranial and teleological version of nineteenth-century phrenology, with (art) modules in the brain taking the place of bumps on the skull, an idea even neuroaestheticians reject.[33] Communication scholar Davi Johnson Thornton sheds light on the parallels between these discourses: "nothing about the idea of location . . . necessarily leads to determinist conclusions. This rhetorical tendency is fed by a long history of localizationist theories that speak in metaphors of hardwiring and machines, which imply static structures. These theories tend to associate location with a sense of fixity, stasis, and natural determination."[34] Recent evidence of this paradoxical and oversimplifying intracranial phrenological approach is given by Zeki's understanding of beauty, an age-old word the neuroscientist considers vague yet deems wise not to eschew, given its widespread use. His argument here is not opportunistic, for he himself acknowledges that "the flexibility is an asset in linguistic discourse"; nevertheless, as he also observes, "the imprecision is a disadvantage for scientific communication."[35] Hence, Zeki acknowledges the classical aesthetic category by name but alters its meaning, inadvertently signaling a commitment to redefining basic aesthetic principles. "Beauty," he claims, "is an experience that correlates quantitatively with neural activity in a specific part of the emotional brain, namely, in the field [A1 of the medial orbitofrontal cortex]; the more intense the declared experience of beauty, the more intense the neural activity there."[36] To support this stance, he invokes Burke's idea that it acts "mechanically" upon the human brain. Such a historically revisionist enthusiasm for an eighteenth-century notion of cerebral mechanisms proves instructive regarding Zeki's systematic devaluation of intellectual context and philosophical tradition.

Zeki's full-blown reductionism is not just methodological; it extends beyond the choice of cases and modes of inquiry. It is prejudicially unconcerned with investigating how we manufacture visual reality. He sweeps away any question of the documentary status of artworks, their stylistic value, and their materiality, from the chemicals involved to the traces left on the canvas by the creative processes via trial and error to the veneer, all of which also affect our enjoyment of them. Nor can he explain significant cases such as those of patients with deficits in visual imagination, studied by Chatterjee, who paint properly when copying models and crudely when seeing with their "mind's eye."[37] Finally, he rules out the "horizon of expectation" that governs our gaze. If Pygmalion is the artist par excellence who blends artifacts and the lifeworld, matter and imagination; the one who breathes life into his own ivory sculpture, with which he has fallen in love; the one by whose kiss it becomes a flesh-and-blood woman, then Zeki is the anti-Pygmalion who certifies optical facts as a matter of fact. Such a critique has been leveled at Zeki from within neuroaesthetics itself and remains in force to this day. Amy Ione stresses the impossible-to-ignore issue that figurative artists touch the canvas with the brush, representing a world they too, as human beings, experience on more than just an optical level, and respecting which they "evaluate [different] hypotheses,"[38] for instance, by choosing to portray bodies by mimesis or deformation. Zeki's analysis, Ione rightly argues, is static. What is more, I would add, it gives rise to an aesthetics of platitude, whose master is the brain. Zeki's take on neuroaesthetics is not only defective with respect to what it overshadows but is also quite productive for what he proposes: a naturalized image for a naturalized beholder who looks at one visual pattern after another, in a vacuum, without interplay with other visual patterns and medial environments. Given my stance, I welcome concerns from complementary perspectives such as that of neuropsychologist Calvo-Merino about similar neuroaesthetic studies on figurative art grounded in the analysis of static and frontal images that are split off from the surroundings and thought of for the most part as mere optical stimuli.[39]

To claim that the brain sees and makes is to lapse into a form of scientifically updated metaphysical thinking. By initiating "an exploration of the relationship between the physiology of visual perception, brain activity and the aesthetic experience of visual art,"[40] Zeki shifts the stakes from anthropocentrism to brainocentrism. His inquiry seems more akin to a logical calculus, or at the very least a rational periphrasis, that in fact anesthetizes the aesthetic qualities of the work itself, from both the artist's and the beholder's points of view. To begin with, he takes for granted that human beings see what they look at as if they were images, whereas field surveys suggest that seeing signs does not lead directly to seeing figures,[41] and art-historical research attests that seeing a pattern is not the same as seeing a picture.[42] What is more, Zeki's perspective is not only ahistorical but completely alien to the dimension of play, and even of conflict, inherent in artistic creation and aesthetic beholding. Pointing to the series of Cézanne's paintings that depict Montagne Sainte-Victoire and emphasizing the role lines play in it (although the lances in the

Renaissance paintings of Paolo Uccello are also lines, as are trees in the forest and lampposts in the street), Ione objects that artistic creation is actually a "dynamic, experiential, and embodied activity."[43] Visual experience itself is indeed far more than just optical. I will later discuss the limits of this exceeding. For now let me simply recall how, in *Late Roman Art Industry*, Riegl taught us that to account for stylistic differences in the history of figurative production, it is necessary to distinguish among an "optical-chromatic" vision of the work, which characterized Byzantine painting and, later, Mondrian; a "normal" or "tactile-optical" vision exemplified by classical Greek art, in which the hand is subordinated to the eye; and a haptic vision, found in Egyptian art but also valid for abstract expressionism and Pollock, in which the eye is subordinated to the hand, whose work and proximity to the canvas it tracks.

In conceiving of artifactual images as extensions of the visual brain, Zeki in fact declines to explore their specific agency. Cognitive scientists Pierre Jacob and Marc Jeannerod are well aware of this; they remind us that the ability to see an image as an image is precisely what distinguishes us from other animals, and that the "make-believe" that underlies beholding in the visual arts is not primarily a neurally based perceptual phenomenon.[44] For, as we will see, even within the neuroaesthetic inquiry, there is no need to suspend the synesthetic quality of visual artifacts, their well-acknowledged affordances, and so on. However, if the brain's task is to achieve knowledge of the world, and the whole point of seeing is to acquire such knowledge by distilling "from the ever-changing information in the visual world only that which is important to represent the permanent, essential characteristics of objects,"[45] then art, as a "product of the activity of the visual brain,"[46] can be nothing but a brain function. This leads to the equation perceiving = thinking, of which art is a by-product whose two "supreme laws" are constancy and abstraction. It follows that the innocent eye and the primitive naivete of the beholder are myths, not because, as Gombrich believed, an untrained eye is unable to see and an untrained hand would not know how to paint, but because, on the contrary, the brain scans reality in an unmediated way. There is nothing but an embrained eye, whereupon the distinction between the Self and the environment breaks down. The work is transparent to the world, which in turn dissolves into it: there is not only no difference between art and artifacts, but no difference between art and living bodies, between artistic and aesthetic. In the end, Narcissus was right not to see his mirror image as an image! Zekian neuroaesthetics sets up a *mise en abyme* of the visible realm where the visual cortex and artists overlap in reproducing the external world and artworks "forcefully"[47] obey the universal laws of the human brain. That is not to say that nature and culture are co-constituted, but that they are the same.

In an essay from 2004, Zeki stretches such an approach into an inquiry about the feeling of romantic love, with the aim of testing how, "exalted though it is, there is a sense in which it, too, obeys a universal rule of brain activity, namely the formation of ideals, a product of the abstractive powers of the brain."[48] A case in point is Dante, who, in describing Beatrice as the "glorious lady of [his] mind" in the *Vita Nuova*, really means that she is his mental construct.[49] This supposed evidence, however, not only ignores any question of narrative code—which makes it nothing short of "ill-formed"[50]—but is not backed by neuroscientific data, if for no other reason than that it is posthumous, and no scholar can provide a postmortem clinical analysis of someone's intentions, especially not by in vivo brain imaging. Quite apart from the presentisms, or the timelessness, implied in Zeki's brand of neuroaesthetics, we would do well to establish what the neuroscientist understands by "ideals." Zeki engages with a Platonic ideal in cerebral terms. We read not only that it can be defined with respect to "the functions and functioning of the brain," but that it is "the brain's stored representation of the essential features" of all the objects it has seen and whose constants it has selected; therefore, the only way to understand the Platonic *eidos* is in reference to brain functions, whose task is precisely that of "[representing] objects as they really are."[51] In *Splendors and Miseries of the Brain: Love, Creativity, and the Quest for Human Happiness* (2008), Zeki refines his argument, defining the ideal as a "synthetic concept," though he does not significantly alter his reasoning, which is not deductive, inductive, or abductive but circular: the brain is made to represent the ideal structure of objects and, *therefore*, the ideal structure of objects can be found in the brain. The neuroscientist dissolves with a metaphysical sleight of hand the polarity of concrete and abstract, sensible and formal, empirical and theoretical, whose groundwork was laid by Plato himself. In establishing the rationale behind the multiplicity of entities that change, decompose, and disappear, such as humans, things, animals, and plants—in short, the world of our experience—Plato conceived of a permanent and stable idea; to this Antisthenes objected that while he saw many horses, he saw no horseness.[52] Conversely, at the beginning of the nineteenth century, when Goethe drew for Schiller an outline of the original plant [*Urpflanz*], taking this to be an idea that can be experienced and is offered to intuition, his friend rebutted: "That is not an observation from experience. That is an idea."[53] What these exchanges illustrate is the dialectic between the contingency of an actual experience, in which we partake at first hand, and the perfectness of a conceptual definition shielded from the ravages of history, time, and personal proclivity. This hybridity is precisely what makes Plato's theory worth considering from an aesthetic point of view. An idea is neither just an efficacious example nor a pure criterion. It is not tangible, but it shapes reality. It is to be grasped by the mind's eye, but it is the matrix of the real. As Cassirer points out, it "is authentically indigenous neither to the world of the sensuous nor to that of the intelligible but, to a certain extent, hovers between them in an undeterminable middle."[54] Zeki, rather, creates a short circuit out of this tension: on the one hand, he neurobiologically materializes the concept of form, idea, constant—hence, idea and brain overlap—and on the other, he universalizes the empirical multiplicity under an all-encompassing abstraction that takes Calder and the falling autumn leaves to be one and the same. What, then, becomes of the subject matter of aesthetics? It meets the same end as a fly seen by a frog's visual system and digested in its belly: it vanishes. If the only way to understand the platonic *eidos* is in reference to brain functions, whose task is precisely "[representing] objects as they *really* are,"[55] it follows that there are objects only qua embrained. This is notwithstanding the caveat voiced by Zeki himself whereby, from a cerebral standpoint, it is not possible to have ideal visual forms without prior exposure to the concrete visual world.

I address the same anti-naturalizing objection to neurobiologist Margaret Livingstone. She argues that artists "have used the assets of our visual system in creating works of art that can be considered outright experiments";[56] in so claiming, she equates the work to the results of a test, which can be either a success or a failure. Kandel endorsed a similar position as late as 2016: "Pollock seems to have grasped intuitively that the visual brain is a pattern-recognition device."[57] It is against this theoretical backdrop that experimental studies like the one conducted by Alumit Ishai and colleagues unfold.[58] Twenty-six subjects who had very little familiarity with the visual arts were called upon to identify the objects represented in sixty works by various artists and in sixty paintings by Robert Pepperell that were similar to them in terms of color, tone, and compositional structure. Each set consisted of thirty color and thirty monochrome paintings. Participants were asked to indicate quickly whether an image contained familiar objects. As it turned out, the subjects perceived recognizable objects much more quickly in paintings that were mimetically faithful than in paintings whose figures were more indeterminate. There is no reason to dispute the results of this experiment. But the fact remains that the analysis reduces paintings to perceptual data processed by the beholder's cerebral mechanisms and interrogates them as stimuli (and not as the source of a beholding experience), to certify that in given conditions, the subject has responded in a given way: the painting is defined as an immediate burst. Beholder and work coincide as if in a lucid dream.

The overlap among artworks, perceptual data, and spectators also runs through the pages of Livingstone's *Vision and Art*, where she distinguishes between a "where" system and a "what" system: the first, which is older and present in the dorsal region of all mammalian brains, is color-blind, is sensitive to small degrees of luminance contrast, is faster, has lower acuity, and recognizes movement, space, position, depth, and figure-and-ground segregation; the second, which is more recent and present only in primates, is located in the temporal lobe of the brain, sees colors, is slower, has higher acuity, and is responsible for recognizing objects, including faces, colors, and details. In theory, we can read using the "what" system alone, although doing so makes the task more strenuous; this is why advertisers exploit this system—for example, by using colors that have an equivalent glow (which creates a sense of motion and vibration) and cannot be seen by the "where" system—to get people to consume the written image more attentively. Here, however, it makes no difference whether we are looking at the colors of an advertisement or the red and blue of Richard Anuszkiewicz's *Plus Reversed* (1960).[59] The point, which Zekian neuroaesthetics misses, is rather "what that which is seen *shows* us."[60] It concerns unmediated opticality, which dispenses with matter and media. In short, it avoids tackling the question of the medial image's agency, all the while putting forward a highly effective theory of the image, which has led to the emergence of what I call an embrained formalism. Architect Harry

Francis Mallgrave offers an example of this when he writes: "Cézanne [exploited the biology of the visual processing system] with emphasis on lines, edges, and geometric forms based on the cone, sphere, and cube. And whereas other painters from earlier periods employed the same techniques, Cézanne's importance to the development of modern art was his *elimination of content* to the point that the colors and simple geometries of his shapes took precedence over any representational values."[61] The very idea that painted cones and cubes do not re-create the external world only holds up against the notion that forms are free from experience, floating in an intangible, nonbodily reality.

However, even Zeki is aware of the plurivocity, if not the equivocity, of visual artforms and of the imaginary of artists. In *Balthus ou la Quête de l'Essentiel* (Balthus or the search for the essential) (1995), he converses with the eponymous painter, a friend of his, dwelling on the psychological force of Vermeer's *Lady Standing at a Virginal*, the magnetism that characterizes the Dutch master's *Lady at the Virginal with a Gentleman* (*The Music Lesson*), and the suspended ambiguity of his interior scenes.[62] Who is the man in the painting? The woman's husband? Her lover? A confidant? The art historian Michael Fried reads this inner concentration in terms of "absorption." The mid-eighteenth-century French painter Chardin was a master of the effect, which he obtained by isolating "one salient detail that functions as a sign of the figure's obliviousness to everything but the operation he or she is intent on performing"[63] and conveying an impression of duration. A sense of sovereign indifference emanates from the scene toward the viewers, who in turn concentrate on the painting, just as the painter had done before them. Zeki, however, translates the interplay among painter, painting, and viewer—the cornerstone of the pictorial fiction—directly into cerebral terms. The painting's greatness derives "from the way in which its technical virtuosity is used to generate ambiguity"—that is, from "the ability to represent simultaneously, on the same canvas, not one but several truths, each one of which has equal validity with the others."[64] And what does it say in the letter that fixes the woman's gaze in Vermeer's *Woman Reading a Letter*? This is open to interpretation and, at the same time, the artist's mastery, "[neurobiologically speaking, is his] capacity to evoke many situations, not one, all with equal validity and hence to cover a 'whole species of situations.'"[65] According to Zeki, what makes this possible is situational constancy—the existence of a permanent variable form of different situations that focuses viewers' attention. This reasoning recalls the insights of Lessing and Goethe about the beholder's imagination. Lessing, in observing the sculptural group *Laocoön and His Sons*, praised the artist's ability to represent the action *before* it reached its climax, thus leaving us with the (exacting) freedom to imagine and complete the action. In the same vein, Goethe claimed that figurative art needed to seize on a "passage from one stage to another"[66] in the represented action, prior to its conclusion, which should be foretold by the body's posture. This dynamic trait proves decisive for our ability as beholders to return to look with emotion at the work—which gives off more than just optical information—by ensuring its reenactment at different moments throughout our life, and at various times throughout cultural history. Goethe's and Lessing's considerations are not rhetorical exercises aimed at safeguarding the eternal value of art but speak to the imperative that the work never be dull. George Kubler elaborates on this in *The Shape of Time* (1962), writing that *déjà vu* and *trop vu* are nothing but forms of boredom, excessive habituation, jaded familiarity—the enemies of any strenuous aesthetic commitment, which alone can give pleasure by bringing the beholder face to face with the art. Changeux himself ventures a playful interpretation of the task of beholding: we turn our head and our gaze, directing them toward stimuli that are new; "randomly distributed neurons in the brain stem . . . help to regulate these movements." He adds: "When a surprising event is repeated, losing its unexpected character, the amplitude of the orientation reaction progressively lessens."[67]

But Zeki's inquiry excludes the temporal dimension, the recurrences and repetitions that have a part in aesthetics. The decisive move in his thinking is to assimilate the incomplete to the ambiguous, whose production and beholding, he argues, do not depend on the special technical skill of the artist, on a mastery of the medium, or on the educated gaze. It would thus be interesting to know his interpretation of Michelangelo's *Moses* and to compare it with Arnheim's observations about the same sculpture. Arnheim stresses that if Moses's angry gaze falls on the Jews, who are imagined to be standing at a remove from the sculptural space, we as viewers must not make the mistake of completing the composition "by supplying the 'missing' distant crowd in [our] imagination," lest we distort the figure the statue actually shows.[68] The completion, as a free play of the imagination of the viewer who beholds the plastic figure and identifies with it, in its past and its future, does not perceptually exhaust the work, contrary to what Zeki's analysis seems to suggest. In a related sense, in "The Neurology of Ambiguity,"[69] Zeki speaks of a gradualism between the maximal degree of the unfinished in Michelangelo's *Rondanini Pietà* and its minimal degree in the Kanizsa triangle, an incomplete triangular figure that the viewer fills in: this, Zeki claims, is "in a sense, a neurological trick"[70] that was used as far back as Cycladic art, whose statues emphasize parts of the body such as lips and noses, allowing our imagination to interpolate the context. By implication, there is no distinction between statues and lines, bodies and depictions; nor is there stylistic specificity: "art must, after all, obey the laws of the brain"[71] since it amplifies the brain's functions and shares its goals. In looking at it, we pass through the looking glass of our eyes.

Surprising though it may seem, Zeki's analysis leaves no place for a real phylogenetic inquiry, whereas many other scholars dealing with neuroaesthetics invoke one. In "Form and Meaning in Art," the cognitive scientist Aage Brandt suggests an epistemological framework for neuroaesthetics, proposing that its object be the legacy of art history in its development, that the way it operates depend on the neuronal process of perception and of the mental organization of the cognitive activity elicited by the work, and that its stakes be evolutionary.[72] Olaf Breidbach, however, draws a firm distinction between an evolutionary aesthetics and a neuronal aesthetics. A scholar of Haeckel—who, in a manner befitting his research, had his brain dissected after death—and of Goethean morphology, as well as the editor of a work on the nervous system of invertebrates, Breidbach develops his inquiry in line with the German aesthesiological psychology of Fechner and Sigmund Exner.[73] It is particularly to the latter that he appeals to emphasize how aesthetic data reside not merely in brain signals but in the brain's interpretation, which evolutionary aesthetics is nevertheless unconcerned with describing. In other words, although evolutionary aesthetics is interested in understanding how certain neural patterns have become established, it does not interrogate what these patterns produce. Neuronal aesthetics, by contrast, aims "to understand what is really meant when we refer to something as being objective,"[74] given that "objective" here refers to a construct of the brain.[75] Breidbach proposes a research program that—in an attempted dialogue with the humanities that also highlights the divergent epistemological perspectives—tracks down the primary brain functions, by examining neurophysiological processes, to determine if and to what extent these also affect the beholder's behavior. In short, neuronal aesthetics endorses the idea that "it is not the outer world we represent in our way of looking at images; rather, it is our own understanding and our own categories through which we establish our world view."[76]

Such an attempt to negotiate between cerebral processes and media production is lacking from Zeki's visual neuroaesthetics, which is epistemically blinded by its dependance on neuro-opticality. Although the brain does not need to be conceived of as a means "to render oneself docile vis-à-vis one's environment, in a word to adapt to everything, to be ready for all adjustments,"[77] Zeki holds it up as an agent, and as "no mere passive chronicler of the external physical reality but an active participant in generating the visual image, according to its own rules and programs."[78] It is symptomatic of the reductionism to which Zeki subscribes to construe such rules as actual laws that cerebral processes obey, leaving no room to imagine how artistic *fabrications* might occur. He equates the material object of the inquiry (artworks) with the inquiry itself (cerebral hypotheses); the painting is aligned with its beholding and the beholder with the painter, who is conceived of as a neurologist studying the brain and its mechanisms, though without being aware of it.[79] In short, Zekian neuroaesthetics is about art production without image-making processes, artworks without agency, and spectatorship without gaze, when really we can espouse a materialist approach based on the idea that living bodies and artifacts coevolve and still reject the notion that cerebral processes have an autonomous epistemic agency. We can be materialist without naturalizing brain activity.

## FABRICATED BRAINS

The second trend in neuroaesthetics I mentioned at the beginning of this chapter is exemplified by the theoretical and clinical work of Vilayanur S. Ramachandran. Like Zeki, he too started out working on visual processes and has pursued a consciously brain-based approach to artworks. He maintains that his ignorance about art history, and particularly Western art history, gives him more freedom to identify transcultural laws of artistic practices and spectatorship, notably by comparing Western and Indian pieces. As a result, he has been criticized, along with Zeki, for proposing an "extravagant generalization about art . . . and then discuss[ing] a small number of examples, which are chosen to *illustrate* the generalization they favour and not to *test* it."[80] Would these scientists, asks philosopher John Hyman, tolerate this procedure in their own subject? We face anew the issue of epistemic quality criteria I raised earlier. By contrast with Zeki, however, Ramachandran takes art as a protean domain to be interrogated without recourse to teleological explanations of artistic and spectatorial behavior. The charges are therefore misguided; he does not treat artworks as evidence for his neurobiological accounts, but rather enlists them as tools for inquiring into our extraordinary performances, which in turn sheds light on our ordinary ones. A modern-day Sherlock Holmes,[81] he investigates art for everything unusual or as yet unknown it has to offer.

His search for commonly agreed-upon, invariant laws of composition is inscribed in a tradition dating back millennia. In the fifth century BCE, Polyclitus wrote a treatise on the matter that served as a paradigm for sculptors. It survives today in three fragments that pertain, respectively, to the symmetry of the parts of the body, to their harmony, and to the modularity of the whole—all criteria for helping the gaze see properly. When Zeuxis needed to depict Helen, he sought permission from the city of Croton's five most beautiful virgins to use them as models for the best features of each. In 1967, Hans Daucher, computerizing a process pioneered by Francis Galton in the 1880s, superimposed the photographs of twenty women to obtain an average face; the result largely erased specific differences and was judged by viewers to be more beautiful than the individual models. The year 1994 saw the release of a softcore interactive video game called *Mind Teazzer* that let players construct an ideal woman by selecting the best attributes from different gorgeous female bodies. Heterogeneous though they may be, these are all variations on the notion that an operational canon underwrites the making and viewing of images of the body.

Regarding visual configurations more broadly, independent of any anthropomorphic traits, Gestalt theory expounds six laws: the principle of proximity, which states that visual elements form an image based on mutual closeness; the principle of similarity, which states that we tend to read elements that are similar to each other—based on their shape or color, for example—together, regardless of their proximity; the principle of good continuation, which states that we prefer continuous lines over abrupt changes and thus tend to read an "x" as an intersecting / and \ and not as a side-by-side > and <; the principle of closure, which states that we prefer closed shapes to open ones; the principle of common fate, which states that two elements that move in a similar manner constitute a unity; and finally, the law of *Prägnanz*—or good figure—which states that, among different configurations of the visual space, the simplest and most stable prevails and thereby governs the relationship between figure and ground. In *Visual Thinking*, Arnheim takes perceptual operations to include "active exploration, selection, grasping of essentials, simplification, abstraction, analysis and synthesis, completion, correction, comparison, problem solving, as well as combining, separating, putting into context."[82]

Ramachandran is convinced that understanding the perceptology of forms aids in better grasping the production and beholding of art.[83] He thus operates at the intersection of gestaltism and the search for canons, suggesting that "artists either consciously or unconsciously deploy certain rules or principles (we call them laws) to titillate the visual areas of the brain."[84] I develop the idea of this titillation not as a trigger for seeing but as an opportunity for gazing. Indeed, argues Ramachandran, to learn about the workings of our brain in its actual visual environment, we need to account for three variables: the "internal logic"—the laws that govern the visual phenomenon; "the evolutionary function"—the reason the laws take the form they do; and the "neural hardware"—the manner in which the brain mediates the laws in question.[85] To make a sphere, for example, we would draw a circle clear in its upper part, for our "visual system has a built-in 'assumption' that the sun is shining from above."[86] As a source of light, the sun would illuminate the convex parts and leave the recesses of the lower half in shadow. These three variables depend, in turn, on the three reasons we have for engaging with art. The first is that internal simulations and mental images train us to deal with, or rehearse us for, life; the second is that we assert our individuality through art;[87] and the third is that recognizing objects is pleasurable to our body. Thus, whereas for Zeki the key question was what *brains* do when they look at images, with Ramachandran it becomes what *we* do when we interact with artworks, how *we* negotiate with them.

In 1999 Ramachandran and Hirstein wrote "The Science of Art: A Neurological Theory of Aesthetic Experience," published in the first of three special issues on neuroaesthetics in the *Journal of Consciousness Studies*. The authors proposed eight "laws of artistic experience": the peak-shift principle, isolation, grouping, contrast, symmetry, perceptual problem solving, abhorrence of unique vantage points, and metaphor.[88] Ramachandran later added two more—the first pertaining to repetition, rhythm, and orderliness, and the second to balance[89]—to make a full decalogue, which, as some have objected, may be better suited to design than to art.[90] This critique confronts us once again with one of the key epistemic tasks within neuroaesthetics: determining the status and the agency of the artwork. Instead of contrasting the notion of the artwork as a social construct (fine arts) with that of the artwork as a prototypical form (beauty) or addressing it as the equivalent of an artifact (thing) or a natural stimulant (palatable stimulus), I suggest treating it as a heuristic device particularly well suited to investigating the aesthesiological process whereby we may develop and elaborate our own sense of agency over and ownership of ourselves as living beings embedded in medial apparatuses. We should regard it as the material agent that allows us to articulate and to queer the epistemic dualism of mind versus body, nature versus culture, interior versus exterior, natural versus artificial, and so on. If we opt for this research perspective, Ramachandran's project may prove highly valuable.

Notwithstanding certain criticisms, I deem his approach consistent with the warning, first sounded years ago by the neuropsychologist Richard Gregory, that perception is not about detecting percepts but about testing *hypotheses*,[91] which are grounded in the negotiation between our neural mechanisms and our epistemic background. While Gregory reiterated his caution to Ramachandran and Hirstein, the laws they devised, in my view, suggest less an inquiry into opticality than a study of sight. They do not entail setting up a brain-world circuit whereby we see the environment discretely owing to its visual aspects, which face the cerebral processes, as in a Zekian mirror loop; instead, they posit "gazing" as a cultural performance codependent on brain abilities. I find the 1978 study by Di Fiore and Di Liberti on stimulation by subliminal images to be an elegant forerunner in this respect. Twenty-one subjects were shown photographic reproductions of *The Accordionist* by Picasso, *States of Mind III: Those Who Stay* by Boccioni, *Interventionist Demonstration in the Piazza del Quirinale* by Balla, *The Persistence of Memory* by Dalí, and *Wreckage of the Shadow* by Magritte, each displayed by tachistoscope for 1/100th of a second, five times in a row. The verbal descriptions, adjective associations, and graphic representations given as feedback accurately matched the five reproductions, which, as the authors point out, do not have a clear figurative content that can be arranged for interpretation. From this, Di Fiore and Di Liberti conclude that Picasso utilized "primitive and preliminary forms of perceiving and *evoking* images" by stimulating "*Gestalt-free ways of seeing*."[92] To claim that the visual processes are neurobiologically based, as no humanist would deny, is not to imply that they are predictable or unequivocal. Visual processes are embrained performative processes; any inquiry into them should examine how neuroanatomical mechanisms shape our cultural agency in its various expressions and what the feedback relationship is that both enhances and changes the aesthetic experience on which they are based.

Grouping, one of the principles in Ramachandran's decalogue, is thus less deterministic than it is descriptive. It works by joining separate pieces into a whole and has an evolutionary rationale, which is to help find objects camouflaged in noisy environments. "There is a synchronization of neuronal spikes of those neurons that fire for different parts of an object and it is this synchrony" that satisfies the limbic system, which governs emotional behavior and the assessment of the emotional significance and potential value of events.[93] Grouping is stimulated by excitation and attention. Let us say we are faced with distinguishing a dog: "at each stage in the hierarchy of processing, when a final solution is achieved—when a part of the dog is identified—there is a reward signal 'a-ha,' a partial 'a-ha,' and a small bias is sent back to earlier stages to facilitate the further binding of the features of the dog."[94] Ludwig Wittgenstein described this in his *Lectures on Aesthetics*: "Would it matter if instead of saying 'This is lovely,' I just said 'Ah!' and smiled?"[95] This "Ah!," like Ramachandran's "a-ha," is an expression of satisfaction—to be distinguished from comfort or pleasure, although these may appear similar. It flags a discovery that amplifies our cognitive embodied experience. According to Ramachandran, artists aim to produce as many a-ha's as possible, and in the greatest possible number of visible areas, by creating a super-stimulus relative to other stimuli at hand. Further evidence for this, we are told, can be observed in Picasso's production, which Ramachandran evokes in its entirety, although it is clear that he is referring specifically to the distorting portraits of faces. Gertrude Stein, whom the artist famously painted, noted how, before the Spanish master, "with the exception of some African sculpture, no one [had] ever tried to express things seen not as one knows them but as they are when one sees them without remembering having looked at them"[96]—that is, in a sense, without imitating them. Nowadays, we prefer to explain this poietic vision in neural terms. Gross, Rolls, and Perrett have detected neurons in the fusiform gyrus of monkeys that fire only when stimulated by the face of a specific subject, such as the monkey's mother. None of these neurons covers the entire face; rather, they belong to a neural network that responds selectively to her face. Within the network for the mother's face, different neurons that are contiguous to one another respond to its different angles (profile, three-quarters, etc.). At the next stage of visual elaboration, Ramachandran finds, another class of neurons is activated; these, which he calls "Picasso neurons" or "master face neurons," respond to a specific face—again, say, that of a mother—viewed from any angle.

The hypothesis is that neurons responsible for a single viewpoint of a face converge on the single master face neuron for "my mother" and thus, conversely, the activation of a single neuron for that viewpoint suffices to activate the "mother" neuron. Ramachandran explains how we would respond to being confronted with two simultaneous and incompatible perspectives of a face: "[we] would activate two individual face cells in parallel in the fusiform and hence the master cell downstream will get a double dose of the activation. If the cell simply adds these two inputs (at least until the cell's response is saturated), the master cell will generate a huge jolt, as if we were seeing a 'super-face.'"[97] Borrowing from Deleuze, we might say "Wow! What a faciality!"[98] Faciality is a by-product of the depicted faces spectators behold, and it exerts an aesthetic dominance over the surrounding visual space. It freezes the whole experience of feeling alive into a still image.[99] Of course, we might object that not everyone "a-ha's" Picasso. This objection, though reasonable, is readily dispelled: the higher centers of the encephalon, Ramachandran theorizes, may censor this visceral satisfaction associated with seeing a highly visual face. This would mean that Zeki is wrong to judge Cubism a "failure" in its attempt to imitate the selection the brain makes by differentiating the essential properties of phenomena.[100] We simply do not appreciate it. More to the point, it would follow that art does not duplicate brain mechanisms, nor is it meant to do so; if anything, it flirts with and satisfies—or dissatisfies—them.

The law most centrally implicated in the entanglement of brain processes, individual aesthetic performances, and artistic agency is the peak-shift principle, which Ramachandran puts to work, once again, examining Picasso's portraits. The principle itself is extrapolated from Niko Tinbergen's ethological study of herring gulls. Herring gull chicks are accustomed to taking food from their mother's beak, which is yellow with a red spot; however, if fed with a long yellow stick (even one not shaped like a beak) marked with three red stripes, the chicks will not only respond in the same way but actually prefer the latter, which is yellower and redder than the mother. Since throughout evolution a yellow tube with a red spot has always been the feeding mother, a tube that is "very red on yellow" constitutes a super-feeding beak. Correspondingly, in twentieth-century art we encounter super-forms, perceptual super-configurations, which we might think of as original forms. Over the centuries, Ramachandran argues, artists have discovered "the figural primitives of our perceptual grammar." He elaborates: "They are tapping into these and creating for [our brains] the equivalent of the long stick with the three stripes for the chick's brain. And what [we] end up with is a Henry Moore or a Picasso."[101]

The cognitive linguist George Lakoff, who discusses the sculpture of Moore—specifically, his *Two Forms*—comparing it with *Mother and Child on a Beach* by Corot, poses a question similar to Ramachandran's: how is it possible to understand Moore's sculpture if we lack information about its content? The answer, he theorizes, involves an embedded structure: "There are neural structures in the sensory-motor system that are 'secondary' in the sense that they are connected neurally to 'primary' neural ensembles that are more directly involved in either perception or movement."[102] The premotor cortex, for example, carries out complex motor actions by way of its connection to the primary motor cortex, which controls simple actions. When this connection between the premotor cortex and the motor cortex is inhibited, the secondary premotor cortex can function as a link in the transmission chain, or "cog," which can still compute complex patterns—allowing for inference formation—and evolve over time.

Lakoff explains: "such patterns [—cogs—] can structure what we see as form in art. . . . Many kinds of cogs have been hypothesized [—for example, image schemas, force-dynamic schemas, and aspect schemas—] and each type corresponds to an aspect of form." The cogs, which "are at once embodied, since they are part of the sensory-motor system, and 'abstract,' since they do not include details," give structure to culture, aided also by conceptual and linguistic metaphors: "[cogs] allow us to have an embodied understanding of the form of abstract art, and metaphors apply to cog structures to provide interpretations for abstract art."[103] Lakoff develops such a hypothesis around the notion of "schema"—from the Greek word for figure—which he, together with Vittorio Gallese (whom we will discuss more extensively in chapter 6), defines as an interactive network that depends on the nature of our body, our brain, and our social and material interactions with the world.[104] The art historian David Freedberg, too, has stressed how rich the notion of schema is for neuroaesthetics, crediting its ability to elicit recursive pathic responses. His reference point is the letter from Nicolas Poussin to his patron, Chantelou, dated November 14, 1647, which echoes the discourse of "modes" proposed by Gioseffo Zarlino in his music-theory text *Harmonic Institutions* from 1553. Poussin writes: "As the modes of the ancients were a composition of several things put together, their variety gave rise to a certain difference of mode whereby one could grasp that each of them retained in itself a *je ne sais quoi* of variety, primarily when all the things that entered into the mix were put together proportionally, whence proceeded a power to move the souls of viewers to various passions."[105] Freedberg understands these modes as providing the "syntax" of the correlations between images and the responses they provoke.[106]

The notion of schema as a salient quality of structures recurs in what Ramachandran calls *rasa*, from the Sanskrit word for "essence." A form, then, corresponds to a stable and prominent material organization. It is, on the conceptual level, what the peak-shift effect is on the aesthetic one. A rat that has to distinguish a square from a rectangle and is trained to choose the latter will respond even more readily to longer and thinner rectangles: it will seize on their rectangularity, which is always relative to that of other forms.[107] *Rasa* would likewise denote the femininity of a feminine statue—for example, of the Chola (eleventh-century) bronze of the goddess Parvathi in Ramachandran's private collection. The statuette is posited as a material instantiation of a hyperbolic female form: its voluminous breasts and ultrathin waist amplify the standard female anatomy, triggering a stronger response from the neural mechanisms that react to those features. A major objection to Ramachandran's understanding here is that he in fact misapplies the peak-shift principle. Hyman contends that "the peak shift towards wider hips and larger breasts would only occur if an inhibitory gradient was created around a female body with slightly narrower hips and smaller breasts than average, and the predicted effect would be that the subject's response would peak at a female body with *very* slightly wider hips and *very* slightly larger breasts than average."[108] And there is nothing slight about the Parvathi statue's proportions. But more incisive is Hyman's charge that "there is no evidence that male spectators who find these sculptures beautiful have innate or learned stereotypes that interact to produce a peak shift in their response to female body-shapes" and therefore that "Ramachandran's theory of art . . . could just as well be a theory about Pamela Anderson."[109] Setting aside that one need not be a horny male to appreciate feminine body shapes, such a claim marks one of the shrewder challenges to the neuroaesthetic approach. Indeed, in stressing the distinction between living and fabricated forms, Hyman establishes the existence of mutually exclusive natural and cultural entities whereby failure to acknowledge the proper, specific status of each item discussed leads to an epistemic fallacy. I have raised the same objection, albeit against Zeki, when he treats the (kinetic) art pieces he addresses as heterogeneous from the surrounding world, only to nullify this specific status immediately afterward by erasing their artifactuality. Yet why should Pamela Anderson be considered less artifactual than the statuette? Why should nature be less nurtured than culture? And what if we were to take art as the nurturer?

According to Ramachandran, art's purpose is indeed "to enhance, transcend, or indeed even to *distort* reality."[110] He hypothesizes that the same neural circuits are activated when we imagine tackling challenging activities, such as hunting or courtship, as when we actually engage in them. More specifically, we simulate the activity with a lower expenditure of energy: "Art may have evolved as a form of virtual reality simulation. . . . [Our ancestors] may have created real images ('art') as 'props' to rehearse real bison hunts or to instruct their children. If so, we could regard art as Nature's own 'virtual reality.'"[111] This hypothesis recalls the one put forth some years earlier by the biologist Richard D. Alexander, who suggested that one of the central features of the human psyche—which comprises "the entire collection of activities and tendencies that make up human mentality"—"is the construction of alternative scenarios as plans, proposals, or contingencies in a manner or form perhaps appropriately termed social-intellectual practice for social interactions and competitions (practice which lacks a prominent physical component)."[112] "This hypothesis of scenario-building," Alexander maintains, "sheds light simultaneously on a collection of human enterprises that have seemed . . . impossible to connect to evolution—such as humour, art, music, myth, religion, drama, literature, and theatre—because they are involved in *surrogate scenario-building*."[113] This goes for art production as well as for beholding, since in both cases we create or re-create environments that are other than the ones in which we live but that select some of their more prominent features. They are not less realistic but hyperrealist milieus. "Virtual," then, in Ramachandran's sense of the term, implies a plexus of coexisting body/brain possibilities that can be differentiated and realized by imaging and making.

A primary instance of virtuality on the neurobiological level is the synesthetic process. In 1812, Georg Tobias Ludwig Sachs provided what is considered the first ever medical report of synesthesia, based on self-observation.[114] A German doctor writing in Latin, he frames his inquiry within the Goethean search for *Urfarben*, or original colors, as expounded in the 1810 *Theory of Colors*. He writes about "colored ideas" conveyed by numbers, days of the week, and time periods in history.[115] Previous accounts of the phenomenon—such as Molyneux's questioning whether a blind person whose vision is fully restored would be able to know a sphere or a cube by sight alone—are better understood as cases probing the role of sensory experience in knowledge acquisition, in the spirit of the Baumgartenian aesthetic inquiry. According to Merleau-Ponty, synesthesia results from the mutual interaction of the senses in response to the structural properties of a body. While especially conspicuous in certain subjects, "synaesthetic perception is the rule, and we are unaware of it only because scientific knowledge shifts the centre of gravity of experience, so that we have unlearned how to see, hear, and generally speaking, feel, in order to deduce, from our bodily organization and the world as the physicist conceives it, what we are to see, hear and feel."[116] As a matter of fact, in blind synesthetic subjects, the application of nonvisual perceptual stimuli, whether tactile or auditory, is correlated with increased activity in areas belonging to the visual cortex, despite these having no apparent reason to be active. Indeed, as Mikel Dufrenne writes, "the world is always present to me as sensible";[117] having a body means possessing a schema of the full range of intersensory coalescences, being able to redefine our current experience, and being vulnerable to a nonhabitual material milieu. Such a coalescence is original with respect to single percepts, which offer channels for organizing situations. This awareness, which French phenomenology explored throughout the twentieth century, is currently the subject of several neuroscientific studies.

Today, "synaesthesia is believed to be influenced by both genetic and environmental factors," although how, precisely, they "interact to create specific associations in specific individuals is poorly understood."[118] In a discussion with Edward Hubbard, Ramachandran credits synesthesia as an authentic perceptual phenomenon that depends not on hallucinations or mnemonic associations but on an overlapping in the brain, localizable in the insula and the angular gyrus.[119] It is notably present in the fetus and later in the immature brain, where the senses are not yet clearly differentiated, although it also occurs in adults. The actual percentage of synesthetes in the population at large is still unknown. Given that abstract shapes, graphemes, and numbers are the most common phenomena to prompt automatic color associations (supposedly the Roman numeral III appears colorless whereas the Arabic numeral 3 generally appears red, except if seen with eccentric viewing, in which case the outline of the number would still be visible but without color association), synesthesia also plays a role in grouping, by ordering similar objects, which all pop out together based on the same "perceived color." Types of synesthesia are known to occur in other combinations of elements as well. They can arise between a sound and the movement of the lips that emit it, between a gesture and a vocalization, and between a visual appearance and a vocalization (so that in describing a small object, the lips contract while saying the word "small"). One type is illustrated by the Kiki-Bouba effect, which Ramachandran studied in an experiment by asking participants to match each of two abstract figures, a star-shaped one and an amoeboid one, with one of two nonwords. Nine out of ten respondents agreed that the first figure was called Kiki and the second Bouba, as if to say that an angle sounded like a "k" and a curve like a "b."[120] The effect can be explained by sound-vision synesthesia, "probably between the tone of vowels and the vertical or horizontal aspect of figures."[121] As Paul Claudel wrote, "the eye hears." From a phenomenological standpoint, synesthesia expresses the co-originality of human beings and the world; it is the expression of an experience in progress that we, qua finite subjects, have of ourselves as matter in relation to other matter that consists not merely of forms to be seen, surfaces to be touched, sounds to be heard, or aromas to be smelled, but of dynamic convergences of these various dimensions.

Although "there is no commonly agreed upon model for how and why synaesthesia emerges,"[122] general consensus holds that it is grounded in "neural plasticity," which unsettles the notion of a circumscribed and self-contained personal identity. The term was coined by the neurophysiologist Jerzy Konorski, one of Pavlov's pupils, who defines it as "[a property] by virtue of which certain permanent functional *transformations* arise in particular systems of neurons as the result of appropriate stimuli or their combination."[123] Synaptic plasticity is first and foremost the process underlying learning,[124] whereby habit formation entails plastic modification.[125] Quite interestingly for us, Konorski points out that "mechanisms of plasticity are to be found only in a normal animal, whereas after it has been anaesthetized and its skull has been opened they vanish completely, and so the experimenter who is tempted to make direct investigations into the activity of the cerebral cortex finds that under his hands it becomes only an excitable organ."[126] In other words, synaptic processes define our personal experience in vivo when we perform in an environment and can be triggered by technologies, from the simplest tools to the most sophisticated devices. Neuroimaging studies show, for example, that touching a braille keyboard activates the visual cortex.[127] Technological superimpositions or substitutions of single sensory modalities in different cerebral areas recalibrate and extend individual abilities. Brain mechanisms and media coevolve. This represents perhaps the most decisive counterproposal to the mirroring loop, whereby the embrained image digests the outer world, for visual experience is in fact media mediated.

Such a coevolution is what underpins the material and medial instantiation of the Self par excellence: the cyborg. This term (a portmanteau of "cybernetic" and "organism") was coined in 1960 to describe a new kind of human being, functionally altered to meet the demands of extraterrestrial habitats: "The Cyborg deliberately incorporates exogenous components extending the self-regulating control function of the organism in order to adapt it to new environments."[128] A half century later, the Eyeborg has emerged as a terrestrial iteration of this project. A color-blind subject fitted with a visual-to-auditory sensory substitution device—a small video sensor placed at eye level that transmits color information to a computer chip—is able, after eight years of continuous use, to "see" colors by perceiving them as sounds.[129] The interplay of brain activity, artifacts, and milieu has a mutually recursive effect whereby, as proposed by philosopher Don Ihde and cognitive anthropologist Lambros Malafouris, "we make things which in turn make us."[130] The prevailing conception of the cyborg is indeed one of an organism interconnected with outside reality by way of implantable, external prostheses. The tools penetrate the skin's boundaries and later become established within. We find it thematized in narratives such as the 1989 science fiction manga *The Ghost in the Shell* and actualized in experiments such as the one connecting Professor Kevin Warwick with an out-of-body implant. In 2002, at the Radcliffe Infirmary, Oxford, the British engineer, who has also, more recently, lamented the poverty of our innate sensorium,[131] embedded a one-hundred-spike array in his forearm, thus linking the nerve fibers in his wrist to an external radio transmitter/receiver device: "The nerve impulses running between brain and hand can thus be 'wiretapped' and the signals copied to the computer. The process also runs in the other direction."[132] His aesthetic agency is extended online. Of course, as the philosopher Andy Clark clarifies, it would be a mistake to think that "the most profound mergers and intimacies must always involve literal penetrations of the skin-bag . . . [while] non penetrative cyborg technology is all around."[133] Yet the idea of coevolution and encounter between the brain and the inorganic artifacts always presupposes their prior heterogeneity, which can thereafter be cognitively overcome because "our most significant technologies are those that allow our *thoughts* to go where no animal thoughts have gone before."[134] This idea presupposes that cognitive extensions requires a dematerialization of the living body. What if we were to disrupt this idea and think in terms of body/brain systems? I would raise the brain-out-of-the-skull to the next level by discussing why sensing the outer world is more than just a metaphor.

The primary device we have at our disposal is our own living body. I am by no means unaware that, in passing from the notion of body part as tool to that of body part as object, we ought to discuss the issue of the commodification of human beings, with its attendant moral dilemmas, from organ donation to surrogate pregnancy. I will, however, leave such considerations aside and focus instead on their aesthetic conditions of possibility. The philosopher Helmuth Plessner, whose works have only just started to receive attention in Anglophone philosophical circles,[135] argues that we are originally and structurally unbalanced qua living and cognitive agents. Our body, he maintains, is not an intransitive object but rather a site of negotiation between the body we have at our disposal and the one we are by feeling its presence. Human self-awareness stems from it. The two are mutually ex-centric, and such a systemic lack of coincidence fosters intellectual processes of self-reflection. The primary sense underlying these dynamics is that of touch. Whereas vision reports our limbs back to us as things among things and flags them as something from the outside—our hands typing on the computer keyboard are as alive to the eye as the keys under our fingers—touch, by contrast, is a means of "self-affection": by touching ourselves we gain a psychophysical awareness of being alive. A highly interesting description of this process is offered by the oft-neglected sensationist writings of the eighteenth-century Abbé de Condillac, who draws upon the example of the statue, an anthropomorphic artifact serving as a human avatar: "Its hand and chest will be distinguished by the feeling of solidity that they send each other, and which necessarily puts them outside each other. Meanwhile, distinguishing its own chest from its own hand, the Statue will find its own I in one as in the other, for it feels itself equally in both, and, whatever other part of its body it touches, will distinguish itself in the same way and find itself equally there."[136] What was in Condillac's case an act of philosophical imagination, or thought experiment, becomes, in Husserl's *Ideas II* and later in Merleau-Ponty's *Eye and Mind*, a phenomenological account. These authors found the transcendental self-constitution of the living body on the specific example of the touching hand that, in touching, feels itself being touched. By moving my right hand onto my left hand, which I see as still, I feel my body as at once acting and acted upon, in an everting process.[137] This awareness is predicated on a perpetual feedback loop fueled by the dynamic of self-estrangement whereby we look at ourselves and acknowledge our own originally self-alienated humanity. This diffracted aesthetic experience is the ground for all subsequent development of our sense of agency and ownership of ourselves.

At the neurobiological level, the most instructive account of this dynamic comes from descriptions of the phantom limb "syndrome," to use Silas Weir Mitchell's term for the perplexing circumstance in which a missing body part is felt as present. The phenomenon has been documented since the Middle Ages. In the nineteenth century, Ambroise Paré, a military surgeon and the first to address it from a medical standpoint, described it as "worthy of wonder."[138] In 1866, Mitchell thematized this marvel—then quite common in the United States owing to factory accidents and Civil War–era amputations—from a first-person point of view in his novel *The Case of George Dedlow*. In 1871, he authored the first scientific report on the experience, which has by now ceased to be a mystery. Today, we know not only that it concerns the feeling of amputated or congenitally absent limbs (arms, legs, feet, hands, phalanges, and even sexual organs) as if they were still—or had always been—there, but also that the whole ensemble of sensations has to do with a rewiring of circuits in the cortex, which contains a miniature map of the human body. A homunculus resides in our brain. In 1917, neurologists Albert Sidney Frankau Leyton and Charles S. Sherrington stimulated different zones of the motor cortex in chimpanzees, orangutans, and a gorilla. Stimulation of specific cortical spots elicited specific movements. In the 1930s, Wilder Penfield and Edwin Boldrey, two neurosurgeons from McGill University in Montreal, electrically stimulated the cortices of 163 human subjects and tracked the correspondences between the brain areas shocked and the muscles that contracted. They then drew the resulting somatosensory map, trying to preserve the localization of the various body parts on the cortical surface. Bearing in mind that each of us has a specific homunculus that corresponds to our individual anatomical development—meaning that the cortical representation of, say, a tennis player's arm would be bigger than that of a nonplayer—the hands and the face, and especially the mouth, typically occupy a disproportionately large space on the map relative to the other parts. Unlike in the Borges story, brain and body do not coincide point to point.[139]

As a leading researcher of the phantom limb syndrome, Ramachandran has significant case studies at his disposal. Tom Sorenson was experiencing a terrible itch on the fingers of a hand he had lost in a car accident. The neuroscientist examined him, blindfolded him, and began stroking him with a Q-tip, asking which part of his body was being touched. Sorenson responded that it was his cheek; but at the same time, it was also his thumb, along with the index, middle, ring, and pinky fingers of his missing hand! The phantom hand was ubiquitous: Sorenson felt the same sensations when touched on his left shoulder. What was causing this sensory confusion, which turned the phantom limb into an extreme form of intrapersonal synesthetic coalescence? The distribution of the various parts of the homunculus does not correspond to the actual distribution of our organs; the face is located near the hand and shoulder, and it is therefore likely that, after the amputation, the part of the cortex in charge of the lost limb, being thus silenced, was overrun by neurons from the adjacent areas. Amputated muscles, joints, and ligaments do not inform the motor control center or the parietal lobe associated with the bodily schema that the limb is missing. Intramodal plasticity occurs: adjacent regions of the somatosensory cortex send inputs into the "mute" region representing the missing limb. Magnetoencephalography (MEG) scans have confirmed that the brain redraws the bodily map. What thus takes precedence is the last mapping of the limb stored in memory, and often its last position, especially when the limb had been immobilized or in a cast for a prolonged period. An engineer from Arkansas who had lost his leg from the knee down disclosed to Ramachandran with some embarrassment that during sex he experienced an orgasm in his penis *and* in his foot. The explanation is astonishingly obvious: in the homunculus the genitals are located near the feet.

But how do we convince ourselves that the missing limb is not there? Aesthetics and art theory have been concerned, since the very beginning, with accounting for our beguilement with bodily appearance in its various guises. So how do we explain what I thematize as the agency of an absence? In 2005, Ramachandran and Jaron Lanier, the pioneer of virtual reality, discussed the mutability of personal identity, as evidenced by certain disorders. A case in point is somatoparaphrenia, in which a person denies ownership of a paralyzed limb after a brain injury, thus also denying the paralysis: "this arm near me in my bed is my brother's" or "it belongs to a corpse, get it off me!" would be common reactions. Lanier showed how the issue could be further complicated by virtual reality.

Let us suppose I am connected, via a simulation device, to a virtual gazelle, so that when I move my arm, one of the gazelle's legs moves. In this way, my bodily schema—my mental map of my body—would be modified through adaptation to a different body. Gushes Ramachandran: "Can we be angels, then?!"[140] Celestial aspirations notwithstanding, the aim of such an experiment is to establish whether it is feasible, by way of fabrication, to draw out the architecture of a preconceptual experience, where brain and body are united in their commitment to disrupt the conventional relation of cause and effect. To help Philip Martinez, a patient whose phantom limb was stuck in an uncomfortable position and causing him severe pain, Ramachandran built the so-called virtual reality box: a lidless cardboard box with a mirror in the center, oriented perpendicularly to Martinez's sternum. The box presented the patient with two openings: one aligning with his right arm and the other with his phantom left arm. Martinez was asked to insert both arms in the box and look at his actual arm reflected in the mirror as he moved it. Via cross-modal (visual and tactile) inputs, the mirroring effect made it seem as if he were moving his left arm along with his right. The phantom limb felt as if it were coming back to life and stretching.

Such a process is not limited to the inner experience of individual humans. The living body is much less private than one might have thought. It has been shown that seeing another person touched (i.e., outside one's body space) induces sensations, as it were, interpersonally; these are elicited solely by way of vision without the need for additional touching of the subject's intact limb[141] or stump:[142] "Mirror-touch synaesthesia occurs when observing touch to humans but not objects . . . and not dummies . . . . It has been linked, using fMRI, to hyper-activity in the somatosensory system in response to observing touch . . . . It . . . is believed to have a developmental rather than acquired origin. However, comparable symptoms may be acquired under certain circumstances. For example, some brain-damaged patients with reduced tactile sensitivity report tactile sensations from observed touch. . . . Such acquired synaesthesia may be the consequence of an unmasking of usually inhibitory pathways and the formation of new synaptic connections caused by the sensory loss associated with amputation."[143] Our experience upon seeing another touched, however, is somehow not the same when we live in our intact body as when we are missing a limb. In the first case, we empathize to a lesser degree—that is, without assimilating the alien touch. In the second case, we embody it in the missing limb. Ramachandran and Brang offer the following as a possible explanation: "The 'null' signal from an intact hand prevents activity in the [mirror neurons system (MNS)] from reaching the threshold for conscious awareness. The amputation removes this null signal, thereby causing the patient to literally feel touch administered to another. It is a sobering thought that the only barrier that stops one from experiencing the sensations of someone else is the skin; remove it and the epistemic barrier between oneself and 'other minds' is also removed."[144] Touch is not simply contact but a graduated pressure that articulates the range of proximity we do or do not experience in our environment; the skin operates as a modulable interface. It is not a container, then, but a threshold that regulates the encounter between our actual bodies and distant others, both living and artifactual. It suggests a fluxus of sensible experience whereby the explanatory polarity of subjective versus objective loses its effectiveness and must be replaced with that of fiction versus facticity. Modern fact-minded people face a new challenge, which has become quite evident. We are to be thought of as original cyborgs and not just as living beings with implanted props. Donna Haraway was right, then, half a century after the coining of the term, to conceive of cyborghood as a cultural, economic, textual, and political figure of selfhood. By abandoning the substantialist, monist view of brainhood, Ramachandran's research comes to support certain interesting materialist studies, such as those by philosophers Gilbert Simondon[145] and Karen Barad, which posit the formation of individuals as the result of an intrarelational dynamic in the theater of the living body embedded in artifactual milieus.

The visual and performance arts interact with and enact this living body, expanding its explanatory efficiency. Susan Broadhurst, noting the scientific findings of Ramachandran and of Helen Neville—who observed how in deaf people parts of the brain's visual system take over the areas responsible for hearing—has questioned whether brain patterns are not also "affected by the unusual and diverse stimuli of . . . art practices" such as, first and foremost, those of the Australian performer Stelarc.[146] In 1982, at Hosei University in Tokyo, Stelarc performed his *Third Hand*, which consists of an electronic prosthesis attached to his right forearm. Clark describes it as follows: "The third hand is controlled by the man, via [electromyography (EMG)] signals detected by electrodes placed on four strategic muscle sites on his legs and abdomen. In effect, the third hand is thus controlled by sending commands to these muscle sites, which act (via the electrodes) as a kind of relay center passing on the messages to the prosthesis. Since these muscle sites are not normally used for hand control, the third hand can be moved quite independently of the other two."[147] In 2000, surgery transposed embodied fictionality into real life. At the Edouard Herriot Hospital in Lyon, Denis Chatelier became the recipient of the first two-hand transplant, four years after his amputation due to a fireworks accident. Following the loss, the patient suffered from phantom limb syndrome. After six months of rehabilitation following the transplant, fMRI scans showed that the somatosensory map of his hand and forearm were returning to the regular proportions found in nonamputees; moreover, he had regained sensitivity in his limbs and was able to grab and manipulate objects.[148] As Ramachandran observes, "*our own body* is a phantom, one that [our] brain has temporarily constructed purely for convenience."[149] The body is neither a *res extensa* nor an organic thing but a situated, lived, conflicting experience. It is a world of possibilities that eschew and exceed the skin/skull boundary and that are entangled with their environment. We therefore might do better to think of it as the site where relationships of inside and outside occur.[150]

To sum up, whereas for Zekian neuroaesthetics the artwork implodes into the percept, the painting implodes into the brain, and the thought implodes into the synapses, for Ramachandran it is our neurobiological system in a state of dynamic equilibrium that transmutes into the work, becoming an artifice in its own right. The brain itself is an embodied artifact, which is not to say that it is a machine.

# 3

# NEUROARTHISTORY

## On Emotions, Matter, and Time

In the fourth century BCE, Apelles, having painted a horse for a competition and found the judges to be corrupt, rounded up living horses and led them to look at all the quadrupeds depicted by the contestants. Upon reaching Apelles's work, the horses whinnied.[1] Art can become nature; it can align color and design and thus be deemed interchangeable with the world it depicts. A rivalry between fictionality and lifeworld underwrites the whole history of image making. By the logic of similar anecdotes, image theorist Gottfried Boehm has come to suggest that the true ambition of pictorial art is to negate itself through maximum artificiality.[2] I have argued that neuroaesthetic arguments largely support and exploit this thesis, albeit without thematizing it. As Seeley aptly observes, however, "the idea that part of what makes an artwork aesthetically interesting is its formal structure, . . . and that what makes the formal structure of an artwork aesthetically interesting is the way it 'resonates' with the operation of the visual system are two largely unexamined assumptions."[3] These assumptions, which we have just seen operating in studies of art production and its beholding from a synchronic perspective, crop up, equally unexamined, from a diachronic one as well. Indeed, some art historians have taken up the latter line of inquiry, and in so doing, radically distanced themselves from Danto's institutional art theory.

Extending the scope of the investigation over a time scale entails placing neuroscientific findings on the brain's behavior at the service of the diverse and successive environmental conditions, both given and contrived, whereby people have looked at, made, and reproduced artifacts, since "the embeddedness of individuals in social patterns and historical contexts generate[s] varied epistemic conditions and create[s] multiple ways of perceiving (and enacting) the world."[4] It likewise entails examining how works taken as cerebral outputs can be related to (and differentiated from) one another, depending on the period and context of their making. Scholars who adopt a neuroscientific approach to art history explore how the brain-based biographies of artists intersect with their works and how the works' aesthetic qualities refer back to the neurological mechanisms of both their makers and their beholders. Nadal and Chatterjee understand such a circularity as inherently human and not generic: "We do not produce and appreciate art because it is in our culture, nor because it is in our nature. We produce and appreciate art because it is in our nature to develop artistic sensitivities and skills that are embedded within a cultural fabric. The flexibility of this system invites the possibilities that each experience of art production or appreciation might be different and even unique."[5] Neuroarthistory tackles the emergence and the development of these possibilities, always on the threshold of a naive reductionism whereby the cerebral agency takes the place of the human one.

## WHEN THE BRAIN LOOKS AT ARTWORKS

The term *neuroarthistory* was coined in 2005 by the art historian John Onians, who sought to answer a question his fellow scholars had failed to pose: "Why do people make art?" Surprised at the omission, Onians turned to neuroscience, taking his cue from the experiment conducted on the portraitist Humphrey Ocean. At the invitation of Stanford University for the project "The Artist's Brain"—part of the broader "Artist's Eye" project sponsored by the Wellcome Trust and organized by John Tchalenko[6]—Ocean underwent an fMRI while drawing the portraits of six faces and six geometric objects from models. The experiment was then conducted on a nonartist control subject with the same cultural background as Ocean. A comparison of the results showed a higher initial activity in the fusiform facial area associated with face recognition for the control subject, and a higher activity in the right medial frontal area for Ocean as the work progressed, indicating a greater cognitive—as opposed to visual—effort.[7] This suggests that artists do not see the world differently but visualize it differently, while also likely drawing on memories of what they have already seen. Onians deemed the experiment useful for making us realize that "to understand an artist, [we have] to understand everything that the artist has seen, looked at, all the works that [he or she has] made"[8]—in other words, everything that has shaped his or her brain. The spectatorial condition, I would stress, is prior to artistic production and depends on neuroscientific scholarship to be known. In 2007, Barbara Maria Stafford proposed, along similar lines, "that art history completely reinvent itself using, in part, the ongoing discoveries of the neurological sciences to do so."[9] Much as Onians had done, she lamented her colleagues' failure to incorporate such insights into their research, further arguing that earlier scholars such as Wölfflin, Riegl, Panofsky, and Kubler deserved to be reconsidered from a neuroscientific standpoint.

The agency of visual patterns in engaging our psychophysical attention had already been pointed out by Wölfflin in *Prolegomena to a Psychology of Architecture* (1886), in which the art historian voiced the need for a close collaboration between his field and that of psychology. Wölfflin, mindful that art history is bound to the chronological sequence along which its objects are distributed, remarks: "One can work exactly only when it is possible to capture the stream of phenomena in fixed forms. Mechanics, for instance, supplies physics with such fixed forms. The humanities still lack any such foundation; it is only in psychology that it can even be sought. Psychology would also enable art history to trace individual events to general principles or laws."[10] Psychology and art history have indeed long been intertwined, at least for the likes of Schmarsow, Jacob Burckhardt, and Aby Warburg, the last of whom coined the term *Pathosformel*,[11] or formula for pathos, to describe the cultural schemata of embodied emotions visualized in figures that recur over the centuries. An emblematic instantiation of *Pathosformeln* is the nymph he recognized in the girl entering from the right in Ghirlandaio's fresco *Birth of St. John the Baptist* in the Tornabuoni Chapel at Santa Maria Novella in Florence (painted between 1480 and 1485) and shown in table 46 of his *Mnemosyne Atlas* in her various recurrences up through the Peasant woman in Settignano photographed by Warburg himself.

Though fundamental to any endeavor seeking to locate the image's pathographic quality and the beholder's emotional responsiveness in the body/brain system, the father of iconology is unacknowledged in the current debate, except by Freedberg. His absence is perplexing if we recall that Gombrich, widely considered a forerunner to neuroarthistory, not only studied Warburg but served for seventeen years as the director of the Warburg Institute (which also hosted Onians), where he worked on an intellectual biography of its founder. Warburg's exclusion is especially hard to fathom given that he was among the first to theorize openly about a physical trace imprinted on brain matter by events. Known as the engram, it stores potential energy capable of reactivation by both the individual and the species and remains imprinted in representational forms as well as in various gestures performed by living bodies. Notwithstanding the trepidations and misgivings stirred up by his aura, perhaps ascribable less to the man himself than to his reception, neuroarthistory's complete neglect of the scholar who strove to explain the embodied survival of images across the centuries remains highly frustrating, not least when others' writings seem unexpectedly to bear his mark. Stafford, for example, writes how, embedded as we are in our individual bodies, yet related to the outside world, to people, to things, and to environments, we respond to such relations via our sensorium, by executing a "repertoire of repeated inner and outer motions thus [constituting] a compressed archive of perceptual primes betraying how the body became associated with emotion-laden forms."[12]

Darwin's 1872 work *The Expression of Emotion in Man and Animals* was a decisive reference for Warburg's inquiry, and it still proves vital to contemporary visual scholars—notably, Freedberg and Horst Bredekamp—studying the embodied nature of the emotional processes. In it, the English naturalist, best known for his work on evolution, expounds a series of principles underlying the externalization of feelings, including the direct action of the nervous system, whose effects we experience as expressive movements, and the force of habit, based on universal and innate, and therefore authentic, recurrences.[13] Darwin's was one of the first scientific books to make use of supporting photographs—specifically, those of Guillaume-Benjamin Duchenne de Boulogne, a doctor at the Salpêtrière who in 1862 published an atlas of eighty-four prints titled *The Mechanism of Human Facial Expression*. These photographed expressions, called "electrophysical figures," reflect specific contractions induced by the electrical stimulation of the facial muscles. Duchenne sought to promote "the art of correctly portraying the expressive lines of the human face," which, as he writes in his Preface, could be thought of as an "orthography of facial expression in movement." Until the end of the nineteenth century, such questions largely fell under the domain of physiognomy, a practical science of the face, emblematized in Le Brun's 1698 *Conférence sur l'expression générale et particulière* (Lecture on general and particular expression), whose spirit lives on in modern-day emoticons and emojis.

The artifactual aspect of the expression of emotions, their being medial and resulting from a process of accurate depiction,[14] stands in contrast to a conception of their visualization on the face as supposedly giving direct access to a person's feelings. Of course, the ability to read facial patterns has proven useful in a range of social and historical contexts, from purchasing healthy and obedient slaves to selecting trustworthy ministers to choosing loyal friends. Even today, the study of "natural emotions" remains relevant from an applications standpoint. After 9/11, the FBI and the CIA repeatedly called on the psychologist Paul Ekman to train agents to spot whether and when subjects were lying under interrogation. Ekman's 1985 book *Telling Lies: Clues to Deceit in the Marketplace, Politics, and Marriage* figures among the reference texts in the FBI's *Law Enforcement Bulletin* of June 6, 2003. As Ekman points out, emotions are special because they are devoid of intention,[15] which is not to say that they are devoid of communicational purpose, but that they are not premeditated or contrived. This idea had been attested to previously by the sixteenth-century science of the passions, conceived of as a "science of the invisible."[16] Visible emotions, or "emotional episodes,"[17] as the affectivity scholar Giovanna Colombetti has more aptly called them, are thus singled out as a hermeneutic tool for reading a supposed inner life. But exactly what ghosts reside at the neurobiological, and thus prediscursive, source of feelings?

This question is raised by the work of James Elkins, who, in *Pictures and Tears*, studies the impact of Mark Rothko's paintings on the beholder, focusing specifically on his fourteen canvases at the Rothko Chapel in Houston and paying close attention to the comments left in the white book by visitors on their way out. Beyond simply trusting emotions—he argues that "tears are the best visible evidence that a person has been deeply moved"[18]—Elkins considers them the direct expression of and testament to the intensity of the beholder's viewing: "the more you look, the more you feel."[19] We thus return—albeit under a different sign—to the crux of Lessing in his *Laocoön* (1766), where he acknowledges the representation of moving expressions in art. But whereas Lessing wrote that the more we look at, the more we think, and that viewers should never be shown the extreme lest they be *constrained* by the sensory impression from which they are meant to emancipate themselves,[20] Elkins argues that the honesty of the innocent and self-transparent feelings is what renders them decisive for a pathic epistemology of the artwork. He is clearly not afraid to spell out the subversive challenge that such an approach poses to art history, which he accuses of hindering any close emotional engagement with the work and confining the viewer to a sleepy and mind-numbing understanding of the picture. Historical knowledge, in his view, proves "insidious, because once it starts to corrode your sense of a picture, there is no stopping it."[21] The question, then, is "does learning kill emotion?"[22] As we have seen, this perspective is generally consistent with the neuroaesthetic approach. Zeki himself, deaf to the anxious I-don't-get-its echoing through contemporary art museums, had already declared a few years before Elkins that if knowledge of Western art were a prerequisite to appreciating Picasso's *Les demoiselles d'Avignon*, most observers would be deterred.[23] Yet many are deterred anyway, because the visual cortex is not and will never be the privileged beneficiary of image making. On the one hand, tears wash away the dialectic between case studies and general laws. On the other, this kind of unmediated and diffuse emotionality dissolves the need for art history and philosophy to serve as a critical discourse concerned with multiplying and diversifying the extent of our aesthetic awareness. All that, it turns out, is really an anesthetic. Far from complaining, I see it rather as the very crux of a reasoning centered on the performance of supposedly unlearned living bodies, however entangled they are in medial environments.

Well before *neuroarthistory* emerged as a distinct label, it was Freedberg whose work proved seminal in rejecting the idea of a cultural-emotional antagonism and studying how the visual arts engage the beholder's response. They do so, he contends, by engaging the body, which is "commanded" by the very body of the image.[24] As far back as 1987, in addressing the definitional criteria for the notion of classicism, Freedberg urged a dialogue with cognitive psychology and a shift away from the culturalist frameworks espoused by the humanities, appealing instead to neurophysiology as a first step toward positing biological and psychological invariances for the beholding of images across periods and customs.[25] Since that time, he has defended three working hypotheses: first, that a scientifically provable link exists between the appearance of images and the viewer's emotional response; second, that emotions can be classified in themselves, independently of the educated pathos that grounds any social culture; and third, that emotions are inextricably linked to the body, as Le Brun was among the first to show. In this way, Freedberg opposes the contextualist approach to art history, on which front he has been echoed by Stafford, who accuses it of atrophying any serious inquiry into the "the role of form, morphology, and structure conditioning what and how we see."[26] The most elegant discussion of the cerebral requirements behind art production and spectatorship is in Damasio's *Descartes' Error*, where the neuroscientist posits the existence of a neurobiological "as if" loop, developed in each of us during childhood or adolescence and primed through adaptation to our environment, which causes a given mental image to become attached, through repeated association, to a surrogate of a body state. There are "neural devices," he theorizes, "that help us feel 'as if' we were having an emotional state, as if the body were being activated and modified [, and] such devices permit us to bypass the body and avoid a slow and energy-consuming process."[27] The prefrontal cortex (specifically, the ventromedial prefrontal cortex) and the amygdala induce the somatosensory cortex to behave as it would if our body were really in the conditions hypothesized (or seen in pictures or on stage). This position, which echoes Ramachandran's and Alexander's, is indebted to the psychologist William James, a frequent reference point for neuroscientists, who in his *Principles of Psychology* (1890) famously argued that we do not cry because we are sad but are sad because we cry; in other words, mental states respond to performed physical changes and "without the body states following on the perception, the latter would be purely cognitive in form, pale, colorless, destitute of emotional warmth."[28] From this we might infer that the external nurturing of emotions is itself embodied; it goes from being a heteronomous impulse to being an inner neurobiological process.

Damasio's approach is highly compatible with a conception of the aesthetic experience as embedded in medial apparatuses, while Freedberg focuses on its neurobiological grounds. His conception of empathy duly exemplifies this precultural process. At a first level of analysis, *empathy* refers to our capacity as individuals to take upon ourselves the emotional state manifested by a person or a landscape via facial, bodily, and spatial layouts. It has two sources. One is eighteenth-century English aesthetics, where Burke described it as a form of "substitution."[29] The other is the aforementioned nineteenth-century German thought. Freedberg makes empathy the cornerstone of his work on the beholder's emotional response to the compositional traits of images and, more specifically, to those features that prompt us to simulate the movements and postures of the depicted bodies. His line of study was inaugurated in 1989 with his magnum opus, *The Power of Images: Studies in the History and Theory of Response*, around what might be termed a "psycho-affective inquiry," whereby he heralded a "reclamation of psychology from history."[30] Its emancipatory aim was to make us "lose our education" and "become 'primitive' and crazy"[31] in order to stop repressing and censoring the emotions elicited by, for instance, Titian's *Venus of Urbino*, which shows a lady lying down, seemingly about to masturbate. This, Freedberg argues, would allow us to recognize the "deep cognitive potential that arises from the relation between the looking . . . and the [figured] material object."[32] To explain such a cognitive and pathic capacity, Freedberg aligns himself with Darwin's underlying hypothesis and argues that "emotions are not only to be understood in terms of geographical and cultural constraints, but rather that there is a fixed relation between the outward expression of emotion and the particular emotion underlying that expression."[33] It is Freedberg's priority "to reintegrate figuration and imitation into reality,"[34] by which he means into the lifeworld. The gap between human beings and Apelles's horses thus gets narrower, as the art historian's goal remains "to lay the phantom of high response to rest."[35]

At a second level of analysis, to enlist a brain-based principle for the efficacy of visual patterns as a criterion for cataloguing the artistic production of the world is to propose an approach in which the beholder incorporates and digests the work and no longer needs to be taught to understand it. This brings us to a neuroscientifically updated version of what phenomenologist Moritz Geiger, in 1928, called "dilettantish" spectatorship, which especially pertains to the mimetic arts: the viewer partakes in the goings-on represented in the work by experiencing an "empathy for its content."[36] The believer is moved by Christ in Giotto's *Crucifixion*, the proletariat by the people in Pellizza da Volpedo's *Fourth Estate*. From the neuroarthistorical standpoint, such empathy is rather induced by the contrived pattern, suggesting either stasis or action, with which the beholder identifies neurobiologically. Paradoxical though it may seem, the two perspectives reinforce each other, since both consider the image a vehicle of unmediated excitement. They are instrumental for religious and political propaganda as well as for brain-centered narratives. Bernard Berenson would say we were witnessing that "exasperation of feelings" he envisaged in 1948 when voicing hostility to the vulgarity of Goya, whose most disturbing creations are, by contrast, what draws Freedberg. It was Berenson's view that Brunelleschi and Masaccio had been shamefully neglected in favor of the Spanish painter on the premise that "the squalid, the sordid, the violent, the bestial, the misshapen, in short . . . low life was the only 'reality.' "[37] This objection is not prudish in nature. It raises the question of what can—and cannot—be figured, of the scope of image making and art production. The "visceral intestinal . . . activities" have "no conceivable, visual shape,"[38] argued the art historian and Renaissance specialist, precisely because they are a mere seeing, a perceptual datum, and not a figurative construction. What we thus lose, it would seem, is an understanding of the critical quality of the work. Whereas Gombrich clashed with Ramachandran, one of the least coarse neuroscientists of art, objecting that his neuroscientist colleague had completely misunderstood the sense of representation in art,[39] Freedberg hints at the importance of "[understanding] the neural bases of what we choose to call art," a category increasingly perceived as weak, elusive, and pompous.[40] I find in this contraposition the same point I raised earlier about the distinction between neuro-opticality and neuro-visuality. Insofar as neuroarthistory, like neuroaesthetics, takes beholding as its ultimate object of inquiry, it marks a disciplinary shift away from the production of the work and toward the range of responses to the work, by dismantling the twofoldness of the artifactual image, its duplicity, its mastery of our gaze. The very notion of art once again comes into question.

The pathetic quality of the image, which is not only felt but assimilated into the matter of the works *and* the bodies of the beholders, is also one of the axes of Changeux's work: the French neuroscientist presents the beholder as calmly yielding before a painting and storing its relevant attributes by a Darwinian evolutionary process that involves brain states without directly affecting the organism's genetic material. The time required for such an evolution supposedly ranges from a tenth of a second to several years. Following this model, when mental processes are in play, different combinations of active neurons are spontaneously and transitorily realized. Changeux calls these "pre-representations" and notes that their distribution varies at random from moment to moment. By way of selection, some combinations are retained while others are discarded: "a pre-representation will be stabilized if it echoes a perceptual representation brought about directly by interaction with the outside world. In the proposed model the latent trace left in the brain by this echo brings about a change in the efficiency of the contacts between nerve cells, or even in the number of those contacts. It is a matter of direct concern to us that this formal network of neurons also stores temporal sequences, or concatenations of representations in the memory. According to this model, which is still quite rudimentary, a painting, or rather the significant relationships present in it, is not stored in the memory in a passive manner, like the print left by a stamp on a piece of wax. If storage is to take place, instead, it requires active participation on the part of the viewer—*alert anticipation*, which is always situated in a definite context in time and space."[41] For its part, then, the picture is not just a meme, defined as a meaningful cultural element such as an idea or a story that lives and reproduces itself among people through imitation, but an extremely complex meme, or even "*a complex synthesis of memes* that are transmitted and propagated by the painter's brain from one painting to another in the work of a given painter and from the work of one painter to another."[42]

Changeux develops his artistic analysis by studying Jacques Bellange's *Pietà* (1612–1616), again with the intent of "pursu[ing] the work initiated by Gombrich[, which could] be achieved by investigating the possible neural origin of aesthetic pleasure and artistic creation, and jointly by reevaluating the evolution of a given painting,"[43] and, like the others, without any reference to Warburg. A collector,[44] curator of multiple art exhibitions, and former chair of the Commission interministérielle pour la conservation du patrimoine artistique national (Interdepartmental committee for the conservation of the national artistic heritage), Changeux retraces the iconographic layers of Bellange's painting, taking as a starting point the work of Irina Linnik—who identifies Parmigianino's etching *Lamentation Over the Dead Christ* as one of Bellange's sources[45]—and that of Kenneth Clark, who in *The Nude* highlights similarities between the Lorrainese painter's theme and Greco-Roman representations of the death of the hero. These he relates to an "oral transmission of memes of form"[46] through simplification and recombination—for example, of Donatello's *Dead Christ Tended by Angels* and of a Gothic pattern of the *Pietà* that shows Christ resting on his mother's lap. After reexamining the perceptual process whereby we approach a painting, Changeux goes on to address the recognition of the work, its "rhythm of forms and figures,"[47] and its specific meaning, which—as he may be the first neuroscientist to date to concede, and so plainly at that—is inseparable from its symbolic and intersubjective value. The fact that a painting is "coded in many ways" engages the viewer's gaze, which, by way of contemplation, has to reconstruct the work and verify the interpretive hypotheses formulated in its regard. Changeux envisions a similar mechanism at work for the collector or art historian tasked with making an attribution. The latter will observe a painting until the schemata it presents appear homologous to schemata previously absorbed and stored: "In that moment, a resonance—a feeling of pleasure—occurs. The resonant internal schema is *selected*. The name of the painter who originated the schema is pronounced. The painting receives an attribution."[48] It is an a-ha moment. Such a process seems to depend largely on the frontal cortex, which plays a role in the attribution of mental states and beliefs, in the genesis of hypotheses and intentions, and in critical thinking. No progress is at stake but a recombination of what has been learned. This dynamic approach also characterizes Changeux's idea that "paintings evolve by *adaptive renewal*"[49] of compositional elements, themes, forms, and figures, without this implying any kind of progress—a piece of hermeneutic wisdom. Whereas for Warburg variation and selection are the morphological principles at work when an artist updates his or her figurative heritage, thereby transforming it, for Changeux these principles intervene at "the highest level of organization of the brain: that of reason." "It therefore seems legitimate," he writes, "to suppose that the frontal lobe plays a major part in the creative process."[50]

Changeux's materialist, neurobiological approach does not entail reductionist naturalization. According to him, the specific quality of the artwork is grounded in this cerebral activity whereby the artist manipulates reality, but the work would hardly leave a trace in beholders' memories were it to resemble a natural object accurately reproduced—that is, if what is depicted the way it is were *already* part of the knowledge stored by the subject. Equivalence with nature would neutralize the work: "The figures that it represents are natural enough to provide semantic classification. But the peculiar artificial character of a work of art, underlined by the painter's style, provides the novelty and the distance necessary for it to be efficiently imprinted in the long-term memory."[51] Starting from these considerations, we could go on to enrich the study of framing by thinking of it as the concrete expression of this *distance*; it is the surprising and unexpected element of the work, then, that triggers in us an otherwise lacking response, habituated as we are to the regularities of our natural and cultural environment. As we have begun to see with my discussion of Ramachandran's thought, the work must disavow, distance, and alienate by creating the paradoxical system of endlessly overturned expectations that is art itself, whereupon we can undertake the adventure of recognition. For the artwork to do so, it has to enact a performative power that is historically informed. To reconcile the dichotomy "between the cognitive and the emotive"—between the domain of perception, hypothesis, knowledge, and truth, on the one hand, and that of satisfaction or perplexity, cold sweats and heartbeats, on the other—which Nelson Goodman denounced as "pretty effectively keep[ing] us from seeing that in aesthetic experience the *emotions function cognitively*,"[52] we need not espouse the presentist perspective of some art historians whereby beholders are treated as contemporary with the dead. It is none other than a neuroscientist who challenges this presumption by upholding the epistemic agency of single works, of the idiosyncratic element that bears an immeasurable degree of heterogeneity and subjectivity, eluding all attempts at neuro-naturalization. We are not Apelles's horses.

## WHEN THE BRAIN LOOKS LIKE ARTWORKS

Onians conceives of neuroarthistory as less a theory than "an approach" for probing the art production of the whole world and endorses an orienting principle whereby we are animals and producing art is part of our nature.[53] In his view, art is a universal *behavior*, a practice whose origins warrant retracing based on a compelling nexus of cause and effect, albeit a less performative and more deterministic one than that suggested by Nadal and Chatterjee. Such a nexus, according to his fixed understanding, is neurobiological, since—as he wrote as early as 1998—"physiology and neurology . . . make possible the sensory, motor and cognitive acts involved in making, looking, commissioning, purchasing, talking and writing that, broadly interpreted, constitute [even] Renaissance aesthetics."[54] More recently, he has claimed that the very contribution of neuroarthistory concerns "the role of outside factors in our mental formation and the role of mental activity of which we are not conscious in governing our actions."[55] What drives his effort to trace and organize the artistic heritage of humanity, as collected in the *Atlas of World Art*, which he directed, is then the operational inflection of an implicitly reductionist ideology. For Onians, three tendencies govern the neural structures of the brain: responding to sensory data in accordance with different environments, imagining and completing that missing portion of data related to given objects, and affecting visual preferences. According to Davis's overview of his method, Onians "urges that the neural routines of vision, including the visual routines of making artifacts, pictures, and artworks, are evolutionary products of visual adaptation to distinct ecologies of visual affordance, such as seasonal illuminations of topography, typical local geology, and the growth or behavior of regional flora and fauna."[56] Onians's vantage point is less that of a museumgoer than of someone looking out on a vast landscape. As a result, his plea for neuroarthistory resonates intimately with works such as those of geographer Jay Appleton, the exponent of habitat theory, who in 1976 wrote that many human-made landscapes—for instance, urban parks—retain the characteristics of the environments most biologically favorable to us, relating "pleasurable sensations in the experience of a place to environmental conditions favourable for biological survival."[57] Onians's "approach," then, downplays the cultural import of his undertaking. He has really developed an entire logic of prediction-in-retrospect, in which he appears to adopt an evolutionary viewpoint while actually looking at the art world as if through a rearview mirror. His investigation considers standard cases—such as the pictorial perspective the Greeks introduced after absorbing a natural landscape with foreshortened angles—but does not try to imagine as yet unknown artistic compositions. In any event, if we wanted to determine which city would best accommodate depictions with parallel lines, Onians would advise us to focus on places already typified by them, such as Florence, where the Arno is straight (as opposed to Venice, with its winding Grand Canal). The views Brunelleschi painted of two of Florence's most angular buildings, the Baptistery and the Palazzo Vecchio,[58] are cited in support of this claim. And yet the Tuscan hills are sinuous, and the Bell Tower of the Venetian Saint Mark's Basilica is a squat parallelepiped. Matthew Rampley is right to ironize about Onians's "crude environmental explanations."[59] Would we be better off understanding this correspondence in terms of what Alberti writes, in his *De Statua* from the fifteenth century, about how the various figurative arts stem from our detecting in the land features that, if transformed slightly and regularly by means of adding and removing, eventually present the likeness sought? How, then, do we gauge the degree of environmental influence?

We are left with the thesis that, owing to neuroplasticity, the artistic forms we develop over time bear the imprint of the specific milieus in which we have been immersed. Thus, for the Greeks, the Doric and Ionic capitals carved around 600 BCE were angular because the coeval landscape was foursquare, and the subject matter of their statuary was the life-size human because their lives depended more on human than on animal labor. These are bold claims, not so much because they refer to the ancient Greeks, whom, Onians observes, are considered a people with a primarily linguistic intelligence,[60] but because for any neuroscientific prediction (no matter how retrospective) to be methodologically viable, it must be testable: we must be able to impose the same stimuli on different subjects at different times under the same experimental conditions. And yet the study of image making and of the related practices of spectatorship is by its very definition alien to these preconditions. As Davis points out, "it is *logically* impossible fully to specify the neural correlates (if any) of visuality specifically as a question of the neuropsychological capacities and adaptations of the visual brain by way of any experiment on living human subjects,"[61] since individual experiences such as commissioning, portraying, or carving depend essentially on social relations. Yet insofar as my caveats are appropriate, they are nevertheless overstated, for the notion of history on which Onians relies is based on temporal succession; it is quite simply tantamount to chronology in time, simultaneity among peoples and habitats, and bodily engagement as passive exposure.

He is nothing if not creative in defending his approach, in some instances testing the limits of its application. This is certainly the case with material evidence, as when he describes the Pantheon's structure as mirroring the morphology of volcanos, which were plentiful in the days of its construction, yet makes the visual analogy clear by juxtaposing a photograph of the Roman building with one of Albano volcanic lake, both taken from above—a viewpoint obviously barred to the Romans of the first century BCE. But it also extends to the handling of scholarly literature, as with his attempt to include visual theorist W. J. T. Mitchell among his scholarly references, based on the latter's suggestion that we abandon the "social construction of the visual field" and rather address "the visual construction of the social field,"[62] when in fact Mitchell proposed the very notion of metapicture to articulate the potential of any picture to double itself, frame another picture in itself, and reflect on itself and its own pictoriality.

After stripping all learned visual practices from beholders and reducing them to a natural state, Onians raises his naturalistic approach from a descriptive to an explanatory level by picturing Jasper Johns in his Manhattan studio, "looking at a large empty canvas and empathizing with his uncle's visceral uncertainties [due to the effects of the Dust Bowl on his fields in South Carolina], [which Johns] had come to share [with his uncle]." Writes Onians of the American painter: "The hand with which he selected and applied paint to the canvas's surface *might have been unconsciously guided by* visual neural networks formed under the impact of 'life and death' fears. Those feelings would have contributed directly to the image's power."[63] That the artistic production was triggered by "unconscious drives" is a general claim we are supposed to take for granted, as was the case with Zeki. However, the idea that instinct guides behaviors is the beginning of a problem, not a solution. Indeed, many Americans witnessed the desolate fields of the Dust Bowl. What kind of reflection took place for one of them to render an image out of these uniform and empty tracts? What manner of neuronal imprinting occurred? Hypothesis verges nearer to suspicion than to supposition. This type of inference is pernicious. A research hypothesis—that an emotionally charged image has been imprinted in Johns's brain (though where, exactly?) and is controlling him—is deployed as if it were a factual argument, and the reference to unconscious processes merely masks the issue, as it did for Zeki.

As I have argued, we are our own first artifact. But what Onians is proposing is less akin to artifactuality than to assemblage. Neuroarthistory, if anything, suggests a neural narrative for the combinatory style par excellence: kitsch (an analysis of which is unfortunately lacking from the *Atlas*). Defined by the art critic Clement Greenberg as the betrayal of folk culture without "any regard for geographical and national cultural boundaries . . . so that the native of China, no less than the South American Indian, the Hindu, no less than the Polynesian, have come to prefer to the products of their native art, magazines covers, rotogravure sections and calendar girls,"[64] kitsch, according to Ramachandran, satisfies the principles of perceptual grouping and peak shift effect,[65] which—in psychologist Colin Martindale's opinion—is a valid heuristic for explaining the evolution of art and fashion, insofar as it leads to an ever greater marking of an aspect or trait we would otherwise become habituated to and end up not seeing.[66] While scholars generally agree that kitsch entails a bold visual mechanics endowed with universal efficacy and, more important, that in it "there is no discontinuity between art and life" since the onlooker "recognizes and sees things in the way in which he recognizes and sees things outside of pictures,"[67] they could not be more divided on the precise stakes of the category—namely, whether it relies on a refusal of cultural appreciation vis-à-vis pictorial artificiality or on a natural comfort with a homogenizing gaze that does not differentiate a spectrum of images. Texts on neuroaesthetics, and especially those on neuroarthistory, broadly propose assimilating neural plasticity to the visual and plastic arts themselves. The brain's conformation embodies artistic styles as well. Now that we know the functions of the two hemispheres, observe Maffei and Fiorentini, "it becomes interesting to ask whether in the broad scope of art history, we can spot manifestations traceable more to one hemisphere than to the other [and] certain considerations may suggest that the answer is yes."[68] These include the fact that Egyptian painting supposedly did not develop according to visuospatial relationships and is rather a narrative of sacred stories and wars, geared toward the brain's verbal left hemisphere. The same principle prevailed in medieval art, the authors contend, but the right hemisphere achieved dominance in Renaissance painting, so that it was not until Cubism that the left regained its rightful place. We are thus supposed to acknowledge an affinity, at least on the stylistic level, between Egyptian painting—taken as a whole, plausibly or not, without distinction between periods—and Cubism, likewise taken as a whole, without distinction between artists. At any rate, such is the case for art production. For beholding, we would have to assume that the brain alternates between hemispheres as it passes from one wing of a museum to the next. Consistently, Onians concludes that if art is a product of nature, its history can only be "natural."[69] He cites the axial symmetry of the human face, marked as a favorable trait in the struggle for survival, as the basis for the symmetry of many artifacts up through the present day. He also cites it as the reason why, thirty thousand years ago, representations of female genitals were more widespread in Western Europe and in southwest France and "the relaxed enjoyment of genitalia would thus have been much easier in the relatively safe and temperate area close to the Atlantic coast than in the colder and more savage continental hinterland."[70] This perspective and others like it, however differentiated, follow the same guiding principles: that all human beings partake in a widespread aesthesiology, that preferences and judgments are manifestations of adaptation, and that beauty is the promise of the reliability of the vital functions that humans perform in the environments in which they evolve. This is a universalizing gesture for art history, which in fact started as a Western discipline and spread in this form to the different countries of the world.[71] The notion of the artwork spread in much the same way. But Onians's gaze, unaware of its own ethnocentricity while looking at "the world as a whole,"[72] cannot but take artworks to be the result of a selection, whether natural or sexual. Yet this attempt to dissolve their opacity—Steven Pinker even thought of them as rivaling "cheesecakes"[73]—leaves unresolved such nontrifling variations as that between a hole, a urinal, and Marcel Duchamp's *Fountain*. We thus find a changed set of stakes for Gombrich's warning that "the innocent eye is a myth."[74] Whereas the art historian, in authoring this statement, meant to stress that perception is historically constructed, for humanists eager to universalize their arguments based on neurobiological findings, history boils down to the struggle for life. Ironically, though, if the eye were innocent, the human species would be extinct.

Evolutionary art theories tend to elide the distinction between understanding and explanation, reflection and expression, by programmatically overlapping nature and culture and pursuing a naturalization of thought. This, after all, is consistent with the thesis of neural plasticity insofar as it is—unnecessarily—construed teleologically and understood as productive. They would find no reason, then, to distinguish the maker of today from the many that came before, were it not for their different responses to different environments, which are always conceived of on the principle of plastic correspondence. It was on this principle that Onians framed his research, thereby founding what I view as a kind of neurogenealogy of artistic production, and a corresponding neurogenealogy of art history that purports to be merely the latest step in the "emergence" of the aid scientific knowledge has offered the study of art for thousands of years.[75] Aristotle and Kant, Alberti and Pater, Wölfflin and Zeki are all key players in this assemblage. Aristotle, Onians claims, applied a biological model to the history of tragedy in the *Poetics* by founding theatrical mimesis on the *instinct* to imitate, and was able to grasp this foundation because, after living in Macedonia, Athens, and Asia Minor, and on the island of Lesbos—all unique environments, rich in animal and plant life—"it was easier for him to see man as just another life form."[76] This would have allowed him to take as natural the fact that "our mental apparatus is shaped by our experiences."[77] The art historian likewise posits that, some two millennia after Aristotle, Winckelmann's neural networks were stimulated by exposure to and repeated direct observation of ancient sculpture, such that they became trained in this art form. According to Onians, the German scholar "had *indeed* unconsciously come to understand one of the neurally based secrets of Greek art's success"[78]—namely, that regular opportunities to observe men nude favored plastic production, and that the sculptors' visual preferences were shaped accordingly. Such a statement may not seem to differ substantially from Malraux's assertion that "men find it easier to agree about women's beauty than about the beauty of a picture, since almost every man has fallen in love, but connoisseurs of painting are relatively few."[79] But it is not simply a matter of stressing the performative quality of our habits, the stylistic constraint of an environment, or passivity as a form of implicit selection in preferences and tastes. On the contrary, the *indeed* emphasized in the claim about Winckelmann contains a tacit epistemic option: the possibility of an art history in the first person, and even of an autographic art history.

Just as Michael Persinger carried out transcranial magnetic stimulation on his own temporal lobes in order to see God,[80] Onians brings details of his own life into the picture. Born in England in 1942, he tells of being eighteen years old and bursting with hormones in 1960 when

*Lady Chatterley's Lover* was published and in 1961 when the Profumo affair (centering on a government minister and a call girl) broke. It was around then that he realized he possessed a body in addition to a soul.[81] In the first place, the insistence on neural plasticity shifts the historical inquiry from the near or distant past to a permanent, homogenous, and homogenizing contemporaneity, within an inclusive and somewhat conservative perspective: things could not go differently, *given the context*. This tends to exclude questions of clash, revolution, and crisis in favor of a continuous present that gathers bygone eras under its protective aegis and allows us to identify with artists from past centuries by reconstructing their life conditions. It is a kind of virtual archeology in which the brain, its functioning, and its cognitive and emotional mechanisms can be paradoxically situated independently of material cultures and discursive practices that are forever lost. I cannot but note a contradiction in a research program that considers the brain dependent on the environment and simultaneously downplays the specific elements that make such a dependence qualitatively significant. At the same time, we cannot forget that this approach is quite authoritative and widespread, for it is in fact largely consistent with the way many institutions engage with art history nowadays. The art historian Hal Foster, commenting in 2007 on the curatorial grouping of works by theme (e.g., "Nude/Action/Body" or "Still Life/Object/Real Life" at the Tate Modern in London), remarked that "this sense of *post-histoire* is, paradoxically enough, a common institutional effect today: we wander through museum spaces as if after the end of time."[82]

In neuroarthistory, scientist, artist, and historian all become primarily and fundamentally beholders; they identify with one another as viewers of the world, and become equals from this standpoint, which is not so much unreasonable as non-self-reflective. Onians argues that the artistic production of the past forty thousand years follows *directly* from our biological nature, which in turn depends on the genetic heritage we share at an essential level with the other living beings on the planet, as evidenced by modern science. Ramachandran would agree: "Some of these universal laws of aesthetics (e.g., symmetry, grouping, peak shift) not only may hold across different human cultures but may even cross the species barrier."[83] The heterogeneity between artworks and behaviors gets reabsorbed, obviating the need to address the vexing question of how "a theory of . . . the history of art, such as the history of painting as an art, may be 'wired into' the visual brain."[84] If such is the case, neuroarthistory starts to become a panhuman visual aesthetics. It discusses artwork as an externalization of the viewer, and it is therefore a history of beholding in art, a science of life *living in art*, that, to impose order on the undeniable passage of time, proceeds by updating its precursors—always with careful attention to the makers' everyday lives. The reference to historians' oft-expressed need for a psychological foundation to their research does not suffice to lend credence to a naturalized historiography. Of course, Gombrich, an indispensable reference for his disciple Onians, collaborated with a neuropsychologist, Richard Gregory, in the editing of *Illusion in Nature and Art* (1973). Later, in the preface to the second edition of *The Sense of Order* (1984), he claimed he wrote the book to establish and test the existence of a "Sense of Order that manifests itself in all styles of design and which I believe to be rooted in man's biological inheritance."[85] Finally, in a 1991 interview with Didier Eribon, he spoke of a "biology of aesthetic pleasure."[86] But if it is true that Gombrich claimed "[his] approach is always biological,"[87] it is no less true that not even he believed the mystery of Raphael would be solved by studying herring gulls,[88] and, pace Onians, Burke would have not "certainly been delighted to learn of the neural basis of [empathy]."[89]

Better informed is the viewpoint espoused by Breidbach, who, along with Karl Clausberg, advocates a "neuronal art history,"[90] taking nineteenth-century German art history and aesthesiology as its main reference points. If for Breidbach neuroarthistory examines artworks, understood not just as the outcome of perceptual and mnestic processes but as the very triggers of these processes,[91] for Clausberg—who advocates a strong interdisciplinary effort—the history of art and of the media offers rich, abundant visual material attesting to the way figurative thought has been conditioned by neuronal processes, which are not invariants of vision but, rather, modes for effectively adapting visual production to changes and swerves. Such an epistemological project is not rooted in the fundamental elision of the difference between seeing and understanding endorsed by Zeki, nor does it have to be. Even Berenson, who fiercely upheld this distinction, readily acknowledged that vision depends on adaptation and readaptation in the human species and that these have given rise to a visual language whose conventions we may share with higher animals. At the same time, however, Berenson assumed a differentiating principle whereby some of these conventions, which may range in duration from seasonal fashions to centuries-old styles, are purely "representational," arguing that "the history of art should be the account of the successive conventions it has suffered up to date."[92] In other words, history is decisive for understanding how the effect of time on our bodies and on the bodies of objects, the forgettings and the rememberings, the violence of humankind and the hazards of nature, the material and intellectual loss due to wars and cataclysms, contribute to the selection of styles through manipulations of matter. Visual truth, I must stress, is a truth in the making: if in Ancient Greece Apelles led horses to whinny at their images, in the nineteenth century painters covered their work in a yellow varnish to give the picture's surface a "gallery tone." Visual appearance is not transparent, immediate, or unique, despite how it may seem in readings that proceed without regard for the becoming of cultures in which we are embodied; it results from the interplay between habit and training, backlashes and blind spots.

Stafford studies this scattered coming-to-appearance by harking back to the age-old debate on art's capacity to make the invisible visible, taking this as the guiding principle of her approach, while appealing with great interest to eighteenth-century British associationism for support. The goal of her research is twofold. She seeks, first, to establish "a cognitive history of images"[93] and, second, to establish "a cognitively inflected way of interrogating the history of images,"[94] in order to answer two fundamental questions: "How does art, especially in its earliest incarnations, enable us to detect the cognitive apparatus of its original creators?" and "What makes people persist in replaying the formal categories handed down to them from a distant past?"[95] These objectives are first and foremost epistemic. To fulfill them, Stafford takes Romanticism as her privileged area of inquiry. The Romantic mind, far from being egocentric and solitary, was bound up with the organic—that is, the environmental and the psychic dimensions of the world. The Romantic mind was embodied. Romanticism, Stafford maintains, was the "first cognitive philosophy of art,"[96] precisely because of its ability to align the universal and the singular, the abstract and the concrete. The figurative art of the period—for example, the oeuvre of Caspar David Friedrich—indeed presents the clarity of an architectural structure, as seen in the skeleton of the tree in *Oak Tree in the Snow* (1829), together with the opacity of the concrete thing represented, the immersion of unity in vastness. To Stafford, this polarity reflects the at once abstract and physical nature that characterizes many of our conscious events, owing to the brain's status as a neuronal network that responds to external stimuli while tending to itself and to its dendrite branches.[97] She expresses the give-and-take between individual and world—and the dynamic energy that informs their relationship—via the notion of schema. As already discussed, a visual schema is a recursive pattern that retains, at each modification, its original relations within what it schematizes. In appealing to this notion (which encompasses Zeki's "Platonism" but not Lakoff's theory of cogs), Stafford aims to integrate biology and culture, which she places on an equal footing. This aim dovetails with a project of closing the gap between form and function.[98] The schema thus serves as the intersubjective grammar ordering manifold givenness, despite being concrete. The patterns of points, lines, surfaces, circles, triangles, and squares that materially govern the harshness of the planet's formation can be understood intuitively because the human brain has developed the same forms. We recognize here the principle of formal purposiveness, which Kant develops in the *Critique of Judgment* and explains in a 1792 letter to Beck as follows: "Nature conforms to our powers of understanding in such a way that, even regarding that which is contingent in nature, we must recognize the existence of laws, albeit only in accordance with that which is subjective in the need of our faculty of knowing (with a view to Judgment)."[99] In this case, however, Kant's transcendental option is neurobiologically validated, and the *need* for knowing is rendered as the *fact* of knowing.

Although the risk, upon adopting a materialist epistemology, of falling into a reductionist perspective is quite high, Stafford herself is committed to avoiding it. Our access to the image, she posits, responds to the Vicoan principle of the *verum factum*, whereby the truth of knowledge coincides with its authoriality: we know the abstract entities of math and geometry to be true because we ourselves create these quantities, and we know the world of cultural artifacts to be true because we ourselves are their makers.[100] As we will see, this proposal is analogous to that of the anthropologist Tim Ingold. We are not mere observers of knowledge and of the cognitive processes whereby we come to produce it: we are actors in it, its key players. Experience and theory are reunited in the human, who is author of both. As I am attempting to make clear, the human being per se is an artist, and such a stance is less flattering than it may seem. By expanding our agency, by considering ourselves *sapiens sapiens* qua *faber* and *aestheticus*, we undermine the privilege of the anthropocentric pedestal, for as makers we are entangled in material and discursive apparatuses that intervene in our cognitive agency and that, given our non-omnipotence, are only partly in our control. This, then, is how I reconcile my own thinking with, for example, Nadal and Chatterjee's statement that "art's universality is the product of our common evolution and shared experience. . . . Art's diversity is the product of our complex, culturally permeable, and flexible brain responding to different environments."[101] This argument, which marks the first step toward developing neuroaesthetics in a performative direction, is aligned with a considerable portion of the contributions I mention.

We could indeed map out a family of scholars around the notion of type and schema, starting with Goethe's morphological contributions. The rationale behind such a grouping would be to stress visual recognition over perceptual stability, as Stafford does, granting objects an "echoic power"[102] to reiterate formal typologies and posit the human being as a "homeostatic system."[103] Her perspective finds supports in Varela and Maturana's conception of autopoiesis, which takes the living being as a system that continuously redefines and reforms itself while seeking balance—in other words, as an integrated network of structural processes that, in regenerating, regenerate the system itself, which is autonomous and subordinates change to the maintenance of its organization.[104] (We should note that Damasio recognizes art itself as having an "effect of homeostatic regulation" that helps restore its vital balance.[105]) In addition to Vico and Varela, Stafford appeals to Johann Gottfried Herder, who in *Outlines of a Philosophy of the History of Man* (1778) discusses the sense of touch: through the hand's "experience, through exploring and trying things out,"[106] we learn the original form of living bodies and artifacts and not their mere external surface recognized by sight. This route, too, is autopoietic. It grants that beholders have the power to interact with the work through their performing and sensing limbs, which constitute a historical variable, and whose manipulation is manifested and reflected in various visual productions. Stafford thus conceives of the phenomenal attributes of a work of art, which express our original aesthetic agency as humans, as responding to explicit neuronal activity. A case in point is paintings, "a special class of images that both coalesce and work to make the viewer coalesce large amounts of novel and taxing information."[107] The distinction between art and artifact is knowingly abolished in favor of patterns that arrest our attention and make us conscious and aware of the world, thus shedding Gombrich's division between the art historian and the historian of visual information.[108] From Stafford we glean that neuroarthistory excludes the historicity of perception from the debate, only to reabsorb it under the notion of neural plasticity. This allows her to advocate an understanding of art as contrived: what has made art *art* is indeed "precisely the material and epistemological difficulty of reshaping what the eye sees,"[109] though at the crux of the debate is the way we, as embodied subjects, present our cognitive abilities directly to the outside through the medium of visual artifacts.

# 4

# NEUROARTCRITICISM

## From the Artist's Lesions to the Artwork and Vice Versa

Within the neurohumanities, medical anamnesis, even ex post, serves as a major heuristic strategy for investigating changes or oddities in people's behaviors. The first subject of this kind of posthumous inquiry was Phineas Gage. The year was 1848: Gage, a skilled foreman on the railroad, had to blast a rock that blocked the laying of new tracks; no sooner had he packed in the blasting powder using an iron rod—three feet seven inches long and one and a quarter inch in diameter—than the powder ignited, sending the rod shooting through his skull, which is preserved as a relic at the Warren Anatomical Museum at Harvard. Witness accounts from the time inform us that Gage did not lose consciousness during the incident and recovered two months later. He maintained full control of his limbs and had no trouble speaking, but his personality had changed: he had gone from being responsible and well-mannered to being defiant and rude. He was also unable to make long-term plans. According to recent skull measurements and postmortem neuroimaging techniques, the damage involved Gage's left and right prefrontal cortex. Hanna Damasio and colleagues cite these findings as evidence of the link between emotion and reasoning.[1] Such an approach to data collection is analogous to the method used in a doctor visit: it serves to make a diagnosis.

The same procedures have been adopted within neuroaesthetics. Just as nineteenth-century retrospective medicine aimed at analyzing, in the light of present-day knowledge, historical cases of hallucinations in distinguished intellectuals of the past,[2] our current understanding of brain mechanisms supports what neuroscientist Jon O. Lauring has called "clinicopathological correlations studies" of well-known deceased artists and their production.[3] The level of interest associated with individual studies tends to correlate with the notoriety of the person concerned. Arthur P. Shimamura has indeed made a case, analogous to that of Gage, for Eadweard Muybridge, a leading figure in nineteenth-century photography. In 1860, Muybridge sustained a neurological injury resulting from a stagecoach accident. He was thrown from the coach, hit his head against a boulder, and was knocked unconscious. Later, in 1874, while Muybridge was on trial for having his wife's lover murdered, his friends and colleagues testified that he had become "irritable, eccentric, a risk-taker and subject to emotional outburst" after the accident and the orbitofrontal damage, whereas before he had been pleasant by nature.[4] Did this change in personality drive him to choose a career in professional photography? We have no way to answer this question.

One could draw up an extensive survey of famous injuries. But what happens when the reports consist of artworks instead of witness accounts? The question of the relationship between the brain and body of the artist and the pattern of the work, between the sense of ownership and agency of the former and the rationale of the latter, has been debated for centuries. Giorgio Vasari tells how the young Titian was "a very healthy man and . . . fortunate," such that it would have been better if in his last years he had not "worked except as a pastime to avoid damaging with less skillful works the reputation he earned in his best years before his natural gifts had begun to decline."[5] Emblematic of this change is *Christ Crowned with Thorns*, which was painted between 1570 and 1576, when the artist died, and is a replica of a much earlier *Christ Crowned with Thorns*, painted between 1542 and 1544. The two works share the same image schema, but whereas in the earlier one the figures are so sharp that the soldier's mail looks sculpted, in the later one the contour dissolves into the color and the brushstrokes are less well defined. As an old man, Titian had weak eyesight. But Vasari also considers the difference between the painter's youthful and mature works from the standpoint of specific compositional properties and observes that the youthful ones were "executed with a certain finesse and incredible care," whereas the later ones were "executed in . . . large and bold brush-strokes and in . . . broad outlines." The difference, he insists, does not detract from the later works or from the possibility of beholding and enjoying them; indeed, it suffices to adopt what we would today call a suitable expository criterion that takes the spectator's point of view into account. If the former works can be viewed "both from close up and from a distance," the latter cannot be viewed except "from a distance."[6] We always approach artworks at a specific site, at a specific time of day, under a specific light, from a specific viewing angle, and from a certain distance, possibly in accordance with its particular style. Antonio Palomino de Castro y Velasco, Velázquez's biographer, tells how the Spanish painter worked on the portrait of Don Adrián Pulido Pareja in 1639 using very long brushes, to keep his distance from the canvas: the picture is unintelligible if seen from close up and "a miracle" if seen from farther away.[7] Wölfflin, in "How One Should Photograph Sculpture" (1896), deplores the haphazard way statues are framed in photographs because a three-dimensional work cannot be looked at from just any angle.

Scholars have always granted that the work responds to the aesthetic conditions underlying the artist's performance. But depending on whether we focus on Titian's myopia or on the pastiness of his late paintings, we end up questioning either the living person or the artifact. This age-old double binary comes through, more or less on purpose, in studies that I propose to class in a subfield of neuroaesthetics: neuroartcriticism. As the art historian Robert Zwijnenberg points out, "when the arts deal with mind-body dilemmas (and this includes neuro-essentialism and its discontents), they do not try to resolve them, but embrace them; they tend to forge new ambiguities, paradoxes, and uncertainties and, in short, function as "a locus of contradictions where opposites may coincide."[8] To what extent that currently occurs we will see in the final two chapters. Neuroartcriticism, for its part, tries to dismantle such new ambiguities, on the ground that "the neuroaestheticians allow us to better understand what transforms a mass of brushstrokes into a masterpiece."[9] Of course, the risk of self-fulfilling prophecy remains high: "The real problem is that when higher functions are investigated, it becomes extremely difficult to find controls that will make it possible to draw conclusions from the experimental results."[10] The same challenge posed by brainhood recurs here, with the artist coming into the foreground as a special case of the Self. Works of neuroartcriticism unwittingly go to fill a dossier of pathologies whose understanding is based on a strong principle of causality whereby only a single interpretation—a neurological one—is valid. This approach drives an aesthetics of total visibility, of daylight. Two main caveats should be kept in mind through following pages. The first is that the artistic examples cited are so well known that their exemplarity may be overstated. The second is that the moment we start treating a work as a biological excretion of its maker, and we subject the latter's entire biography to reduction and dissection, we risk losing track of the very thing we were most interested in learning about: art production. I propose to differentiate two strands of this subfield: a neurocriticism of the artist, grounded in the idea that biographies are embrained; and a neurocriticism of the artwork, grounded in the idea that they function as expressions of cerebral impairments.

## EMBRAINED BIOGRAPHIES

The inquiry into the effects of neurobiological (pre)conditions on artistic production relies mostly on studies of brain-damage localization.[11] In *Cervello e pittura* (Brain and painting), a major contribution to this line of study, Anna Mazzucchi, Giovanni Pesci, and Dario Trento note that painters with damage to the left side of the brain tend to work their way along the canvas, completing one portion of space at a time in a modular fashion, and struggle greatly with rendering three-dimensionality and perspective. Given the medical histories of the artists, works such as those by Zlatio Boiadjiev, Afro Basaldella, Gian Franco Fasce, and Renzo Schirolli could be instructive regarding the effects of left-hemisphere lesions; works such as those by Lovis Corinth, Otto Dix, Anton Räderscheidt, Guglielmo Lusignoli, and Federico Fellini could be instructive regarding the effects of right-hemisphere lesions. The postlesion paintings by Dix and Lusignoli are indeed characterized by an absence of movement and by empty spaces. Painters with damage to the right side produce "incomplete" images, which nevertheless occupy the entire canvas as though it were a *unicum*; these images are "phenomena that might be explained in terms of the *predominantly analytic powers ascribed to the left hemisphere and of the predominantly synthetic powers ascribed to the right hemisphere*," which have become ungoverned.[12] Also common to a greater or lesser degree in cases of right-hemisphere lesion are constructional apraxia (the inability to plan and coordinate movements consistent with finalized motor acts) and unilateral spatial neglect (the tendency to overlook the sensory information coming from the space to the right or the left of the subject's body, on the contralateral side to the lesion), although the neglect may pertain to an external object, to the body of the observer (egocentric neglect), or to the surrounding environment.[13] Even if we now know that the brain's plasticity is a more decisive factor than the difference between hemispheres in determining our intellectual capacities and behaviors, it is tempting to appeal to pathological change to explain phenomena such as the "curtain effect" in Corinth's work, on the left side of the canvas, whereby the colored surface is covered in smudges, producing a two-dimensional result. The neurologist Olaf Blanke suggests that "with respect to the art of Corinth," one can gain a better understanding of his style by studying his disorders and drawing on neurological findings.[14]

Also characteristic of both groups is perseveration: the automatic tendency to repeat the same stylistic feature in one work after another (as in the case of Afro) and to return, within each work, to themes and pictorial structures conceived before the illness (although a thorough iconographic investigation of the artist's entire production would need to be conducted to assess the greater or lesser degree of pathology associated with such an iteration). Finally, both typologies of painters are characterized by an impoverishment in the use of colors, but whereas the former prefer light hues, suggesting calm, the latter juxtapose primary colors to jarring effect and use white to approximate light and black to suggest depth. It should be noted that Mazzucchi, Pesci, and Trento claim to detect signs of ongoing degenerative processes even in the prelesional productions of these artists. A case in point is Afro's *Arancio ceruleo* (Cerulean orange), painted in 1970—a year before he suffered brain damage—in which the figures appear simplified. But without the possibility of corroborating evidence, this reasoning seems forced: can we be certain that *Arancio ceruleo* would not have been the same if Afro had not suffered a stroke? In right-handed painters who have suffered damage to the right hemisphere, the left hand, despite having gone unutilized prior to the lesion and being now impaired, proves—with proper training—to be more reliable for painting than the right hand, which is contralateral and thus unimpaired for the performance of common motor tasks; it is as though the problem "were not so much mechanical" and consisted rather in "translating the compositional thought into its pictorial counterpart."[15] Corrective strategies in such cases vary. Afro, being right-handed, devised the solution of having his left hand accompany his right, with the latter taking the former as a "passive extension" of itself, an always already embodied prosthesis. Lusignoli, who suffered a right-hemisphere lesion and left-side spatial neglect, tried to verticalize the figure or rotate the canvas by 180 degrees to complete it. When Fellini, who suffered a stroke in the right hemisphere, was asked to mark the central point on a line he had drawn, he placed it to the right, as doctors expected, but also decorated the whole space-to-be-filled with drawings and cartoons; he manifested a "productive" attitude, whose seeming purpose was to counteract the constraint.[16]

Thus, "certain permanent functional transformations arise in particular systems of neurons as the result of appropriate stimuli" (this was the rationale behind the experiment conducted on the portraitist Humphrey Ocean mentioned in chapter 3) and cease as the result of cerebral changes.[17] Given the link between active areas of the brain and pictorial output, neuroartcriticism points us once again toward the conclusion that the painting is the brain. This must be what the art teacher Betty Edwards thought when—after finding to her surprise that her students were unable to correctly reproduce a model despite claiming to have a proper view of it—she turned for inspiration to the early research, by Roger W. Sperry in 1968, on the functioning of the two cerebral hemispheres. From there, Edwards devised a teaching method that exclusively targeted the right hemisphere, using exercises that were analytic, symbolic, or logical in nature and therefore hard for the left hemisphere to perform. Her method taught students to see according to the "five basic skills" required by the brain, which are "not only about learning to draw . . . the true subject is perception," and which revolve around the ability to perceive five elements: edges, spaces, relationships, light and shadow, and lastly, the whole or gestalt.[18] For students' skills in graphic rendering, the results were optimal.

The assessment of ongoing degenerative processes in artists, even without reliable diagnoses, can also be a component of neuroartcriticism, and can lead to the complete neglect of the artists' broader biographies. Among the poster children of this approach are Goya and Van Gogh. Robert Hughes, the leading biographer of the Spanish painter, tells how in 1792 he was working on his tapestry cartoons for the Royal Tapestry Factory—having already painted about sixty of them, large-scale and in color, since starting in 1775—when he suddenly left Madrid for the south of Spain. There is no further record of him until January 1793, when he wrote of a severe ailment from which he had nearly died. He described hearing loud noises and buzzing in his head and going deaf; his balance was impaired. He was forty-seven years old at the time. The cause of the illness remains a mystery. Yet, as would befit a fictional anatomopathology, it has been speculated that Goya's auditory vertigo was caused by Ménière's disease, with secondary hypotheses pointing to botulism, polio, hepatitis, labyrinthitis, and possibly even syphilis, without ruling out meningitis. The fact remains that Goya recovered from all but the hearing loss, returned to Madrid, and accepted the commission to paint the frescoes at the Church of St. Anthony on the Manzanares River, where he needed to climb a staircase to reach the scaffolding. He made the climb successfully, and "none of the strain of working high in the dome shows in the completed work—only a radiant and diaphanous *joie de vivre*."[19] He went on to live a long life and produce masterpieces such as *The Nude Maja* and *The Clothed Maja*. Possibly even more unrestricted are the diagnostic hypotheses about Van Gogh:[20] in addition to suffering from gastrointestinal pains, exacerbated by his penchant for turpentine and absinthe—which, in the composition available at the time, had the potential to cause visual and auditory hallucinations, insomnia, tremors, paralysis, and epileptic seizures[21]—he reportedly suffered from epilepsy, schizophrenia, and drug addiction, not to mention bipolar disorder, delirium tremens, acute intermittent porphyria (a metabolic disorder), syphilis, and Ménière's disease to boot (although we should note that at least some doubts have been raised as to the Dutch painter's epilepsy). Instead of drawing a taxonomy of possible diseases, however, it would be more fruitful to assess the performative value of pathology itself. As Antonin Artaud sharply asserted, Van Gogh, in the end, did not commit suicide because he was mad or sick but because he had "bodily become the field of a problem that the iniquitous spirit of mankind has debated since the beginning of time, the predominance of flesh over spirit, or body over flesh, or the mind over one or the other."[22]

Notwithstanding disease's aptness for resolving the mind/body dualism in favor of a cerebral monism, the idea of disease as a marked symptom allowing access to the inner and otherwise inaccessible psyche of artists is widely held, not just among the acolytes of image making but also among artists and art lovers. Flaubert is known for certain to have suffered from epilepsy, although he kept his affliction secret, while Dostoevsky disclosed his own experience with the disease in the pages of *The Idiot*, through the accounts of Prince Myshkin's seizures. Just before the first full-blown seizure, which—we read in the novel—was unbearable, his brain lit up and he found himself in a state of the utmost joy and calm. "What does it matter if it's an illness, then," muses Dostoevsky's protagonist, "what does it matter if it's an abnormal tension, if the result itself . . . turns out to be the highest degree of harmony and beauty?"[23] The reference is now standard, to the point that the medical literature refers to temporal-lobe epilepsy with ecstatic seizures as "Dostoevsky epilepsy."[24] In addition to the seventy-two "epileptics of worth and fame"—one of whom is Socrates—collected by William G. Lennox in his extensive study on the disorder,[25] Franco Magnani deserves our attention. He probably suffered from the same syndrome as Dostoevsky's hero. Oliver Sacks, who met him in San Francisco in 1988, tells the tale of the Italian American cook-turned-artist in "The Landscape of His Dreams."[26] In 1965, Magnani was struck with an unknown brain sickness. Every night he experienced extremely vivid dreams of Pontito, his native village in Tuscany. Despite not wishing to return there, he began painting for the first time, taking the village as his obsessive subject matter. Thanks to an extraordinary photographic memory, which he had possessed since birth, he was able to remember the place down to its minutest details. What is significant about Magnani's case is that his pictorial performance began after the onset of his condition. According to Chatterjee, sometimes brain damage actually stimulates the subject's visual production; for instance, a person with frontotemporal dementia may develop problems with language, attention, and decision making yet start to produce "art" for the first time.[27]

In any event, pictorial performance can be beneficial for counteracting illness.[28] This was the case for Willem de Kooning, who, despite suffering from Alzheimer's disease, thus being deprived of his memory (as occurs in dreams) and given to spontaneous and advanced motor automatisms (which is also characteristic of dreams), painted balanced compositions all his life.[29] According to the neuroscientist J. Allan Hobson and the art historian Hellmut Wohl, de Kooning had needed his frontal lobes to learn to paint, but "this ability," once learned, "was relegated to the subcortical brain, where automatic motor activity and the visual images of art are integrated with emotion, just as happens in dreams."[30] The dreaming brain indeed shows greater activation of the cerebral areas associated with emotion and also generates complex visual images, while the cerebral areas governing memory, thought command, and self-awareness are less active. According to Julien Bogousslavsky, if his work "kept the same quality or sophistication standards, these may in fact reflect some progressive impoverishment of his brain abilities." "Nevertheless," he states, "the indisputable aesthetic value of [his] late productions still supports the point that creativity in dementia may be maintained even while being transformed or displaced."[31] Sebastian J. Crutch and Martin N. Rossor further observe about de Kooning: "An increase in the scientific understanding of the neural processes involved in the artistic process is by no means the only motivation for studying the paintings of individuals with degenerative conditions such as [Alzheimer's disease]. On the contrary, painting can provide a means by which those with dementia who are blessed with artistic abilities can express their reactions to their circumstances and can provide for the viewer a powerful insight into both the disease and how we ourselves might respond to the experience of dementia."[32] The notion that art can improve after neurobiological disease lends support to the idea that "the brain does not harbor a single art module"[33] and reinforces the "increasingly popular" argument "that the arts are valuable in education because of the beneficial effect they have on the brains of those who pursue them."[34] Artworks function as brain food and as health exams. This understanding supports the brainocentrism detailed in chapter 1 and the idea that brain conditions directly dictate our personality.

Continuing along these lines, we are but a short step away from an image theory based on a conjectural and all-inclusive etiology open to every painter. Does the thinning of the bodies in El Greco's paintings *depend* on an alteration of his own body image—in the opposite direction of Botero![35]—such as we also see in certain cases of anorexia nervosa?[36] Or does the style of his figures have something to do with astigmatism, as has also been argued?[37] Did Bacon suffer from body dysmorphia, a disorder in which a part of the anatomy is seen as deformed? This interpretive trend, whereby the compositional structure and the poetics of a work finally become clear only today, thanks to neurological knowledge, constitutes a voyeuristic dive into the life of deceased artists that makes it possible to establish a retrospective teleology. Neuroartcriticism places artworks on the anatomist's table to explain that the result could not have been otherwise. We read page after page about Frida Kahlo, about her passionate love for Diego Rivera, and about the many surgeries she endured, all of which is deemed interesting in view of the polio that the Mexican painter had contracted in childhood.[38] This kind of focus is rather paradoxical if we consider that Kahlo is an exquisite example of cyborg, which obviates any need to speculate about how her natural body might have been. Nevertheless, Philip Sandblom opens *Creativity and Disease* with a discussion about her, to make the case for a link between art and experience, primarily but not exclusively in the context of realistic art. In this connection, it would not be difficult to imagine a neurologically based metacritique. To give an example: Georges Albert Édouard Brutus Gilles de la Tourette invented the gait footprint technique, which consisted in having two subjects with the same motor pathology (hemiplegia)—one suspected of being nonorganic—walk on a sheet of paper laid out on the floor, with the soles of their feet painted to leave imprints. The theory was that heterogeneous pathologies would result in distinct traces. The idea was to uncover (dis)simulation by recording the trace. We could go further to imagine the lively part of Papageno in Mozart's *The Magic Flute* as the musical imprint of the composer's own supposed syndrome, which may have caused him spastic tics, specifically of the face.

What does it mean for the brain condition(s) of a sick artist to be embedded in his or her work? Does it mean that the disease helps free the forces of imagination, as Sandblom argues? Where are we supposed to draw the line between ordinary health, idiosyncrasy, and artistic performance? In short, where do neurological inquiries lead and what do they achieve on an epistemic and on a heuristic level? In effect, they create a medical case history of art; they provide material for a neurocognitive inquiry that studies art and the brain as if both were subject to the same rules. This is indeed the belief held by the cognitive psychologist Robert Solso, for whom "art is, after all, physical material that affects a physical eye and conscious brain."[39] Solso also makes use of contributions from evolutionary psychology and argues that when the brain can "visualize and produce nonverbal abstract representations of thoughts," it can produce art.[40] Finally, in his view, while art itself may be a rather recent notion developed by Western culture, humans have produced it since their very first visualizing performances in the seventy-five-thousand-year-old Blombos Cave. For Solso, even aesthetic judgments, regardless of whether they concern architecture, plants, faces, or objects, depend on the evolution of the brain, which develops to gain a better command over our actions and to find a salubrious environment, good food, and good sex while avoiding anything harmful. He therefore maintains that the networks of aesthetic cells are innate, pragmatic, universal, sexual, and engaged in social activities. He argues that *Cossacks*, which Kandinsky painted between 1910 and 1911, is akin to an "artistic Rorschach test."[41] Bogousslavsky would agree with Solso. He asks how the evolution of twentieth-century art "corresponded to specific changes in mental processes" and why abstract art, which supposedly originated with Kandinsky, "subsequently became one of the major ways of expression in painting in the last century."[42] So much the worse for Apelles, who went to Rhodes in the fourth century BCE to meet Protogenes and, after failing to find him, drew an extremely fine line on a canvas left out in the open. On coming back, Protogenes saw the line and drew an even finer one inside it, using a different color. When Apelles returned, he sensed that he had been beaten, but crossed the two lines using a third color. Protogenes then recognized his contemporary as the master and made no further alterations to the painting: "Hanging among the outstanding masterpieces by many artists it looked blank. For this reason, it attracted notice and was more celebrated than any other work on display."[43] "*Nulla dies sine linea*," it was said of Apelles, who indeed stressed this element as the foundation of the image well before 1926, the year of Kandinsky's *Point and Line to Plane*.[44]

Neurocriticism of the artist erases the distinction between personal cerebral mechanisms and public work without the slightest critical discussion of the meaning of the notion of *art* itself, at least in the twentieth century, which is the de facto privileged period for neurological analyses of artistic production. Is it about technique? Is it about creativity? This strand of neuroartcriticism treats the drawings made by Edwards's students, the Rorschach inkblots, and Kandinsky's paintings interchangeably and the "work" as if it were a behavioral liability of its maker, akin to a drunk-driving record. Cut off from any situated stylistic perspective, the pathological condition is posited as disembodied, without regard for the impact that a given disorder, with its attendant pain and discomfort, has on one's prehensile relationship with the environment and that one's experience of life may, on that account, be both more vivid and less free, forcing the artist to experiment anew with figuration through a revised proprioception (defined as "the sense of static position and movement of limbs and body," independent of vision).[45] Otherwise, why do different artists with similar disorders work not only despite the illness that constrains them but *through* it and achieve heterogeneous results? Methodological concerns notwithstanding, under the guise of neurologically tailored diagnostic hypotheses, the romantic affinity between genius and madness is intact enough to serve as the common denominator for the most diverse array of artists. In *Touched with Fire: Manic-Depressive Illness and the Artistic Temperament*, Kay Redfield Jamison lists a vast number of likely sufferers of cyclothymia, subdividing them into categories based on artistic genre—poetry, prose, classical and non-classical music, plastic and figurative arts—and ordering the contents of each category alphabetically: from Antonin Artaud to Walt Whitman, from Hans Christian Andersen to Émile Zola.[46] In 1995, Sacks complained about "an obsession, almost an industry" of neurological studies on the syndromes of artists.[47] Biography, here, becomes chronicle. What gets overlooked is that the history of an individual pathology, cerebral or otherwise, varies depending on the country, the period, the gender, and the discursive economy in question, as the philosophers Georges Canguilhem and Michel Foucault taught us.

Positing the artist as a synthesis of his or her actual visual perception and performances, to be explained through cerebral impairments, and reading the artwork as their outcome, means developing a biography *in* the work. If we wanted to proceed on this path, we would have to assume that image making is nothing but the externalization of cerebral states. This move is not without consequences: it means giving up the medial function that the artwork fulfills by negotiating among the agency of the body, that of the imagination, and the resistance of the materials. The work, then, becomes the unmediated visual expression of an energetic conatus, and it is most significant when least formalized by cultural performances. Exemplary, then, is the production of Nadia, a severely autistic girl studied by Lorna Selfe, who had exceptional pictorial skills; at age three and a half, she could draw extremely realistic horses in perspective, in periods of under a minute, despite never having learned to draw step by step and without copying her drawings from a model.[48] Consistent with the idea that visual artifacts constitute cerebral evidence, Maffei and Fiorentini detect similarities between the drawings made by blind children and the drawings made by people whose figurative representations "were not aiming to depict reality as they saw it." But was Hieronymus Bosch trying to do so when he painted the imaginary world of chimerical figures in the triptych *The Garden of Earthly Delights*, in vivid, oneiric detail? And was Lucio Fontana aiming to do so five centuries later by slashing—once, twice, thrice, or more times—canvases large and small, white and red? Whatever the case, Maffei and Fiorentini report the example of a young boy who drew two rows of trees, one on either side of a street, as though one of the two were reflected in water and its reflection produced the second row, with the result being something akin to an Egyptian image of a pool surrounded by a garden with trees seemingly "propped sideways" along its edges.[49] More recently, Amedi and colleagues tell us that "congenitally blind individuals are able to create mental sensory constructs and manipulate them. . . . Subject E. A. is a profoundly blind painter who lost his vision at a very young age. Like other blind individuals, E. A. can capture the external world by touch. However, thanks to his extraordinary drawing ability, he can reveal the nature of his internal representation of objects captured by touch in a manner that is unequivocally understandable by a sighted person. Thus, by revealing his drawing to us, he is in a sense revealing the internal representations of his brain."[50] Drawing is not necessarily a visual art. To make this case, Maffei and Fiorentini cite certain pictures produced by blind children, ages ten to eleven, that represent complete human bodies, albeit with the layout of the parts of the figure improperly oriented. Their productions are, then, "the expression of innate schemas and models or those acquired through experiences that are not necessarily visual."[51] The neurocognitive point of view not only treats the painting as though it were the brain but conceives of it as happening in the brain. Yet if it has no need to be mastered, it should be considered not an artlike production but a pure visual performance, a kind of peculiar cerebral sport.

Such a research perspective could underwrite a reductionist approach akin to the one suggested by the studies on lesions and fuel the search for an original figurative principle particular to humans qua living beings. There is historical precedent in the path undertaken at the beginning of the twentieth century with the study of the art of the mentally ill, the art of children, and primitive art. As Rudolf Arnheim commented, it is "as though we are afforded a glimpse at the human mind 'as such' "; in their productions, "what we seem to face . . . [are] the manifestations of an art that rises from the very roots of human perceiving and thinking."[52] Among the first to broach the subject was Marcel Réja, who, in *L'Art chez les fous* (Art of the insane) (1907), argued that men of genius are far more extraordinary than madmen because the former are able to show the forms of the spirit with beauty while the latter naively unveil its mechanisms. Réja was followed by Hans Prinzhorn, who, in *Artistry of the Mentally Ill* (1922), took care to stipulate that pathology in itself is *not* a privileged and sure way of gaining access to art. Prinzhorn—a doctor and psychiatrist at the Heidelberg clinic since 1918, as well as a friend of Ludwig Klages, with whom he shared the view that expression, including graphic expression, was a psychophysical manifestation of human beings—collected and catalogued, between 1919 and 1921, the plastic and figurative works of 170 people with mental disorders (his collection includes 4500 works by 435 patients). Without emphasizing psychiatry and aesthetics as fields, he examines the "*Gestaltung* process," whose essence, he claims, resides "in the most sovereign drawing by Rembrandt as in the most miserable daubing by a paralytic."[53] The *Gestaltung* is the visual emergence of a psychic posture. In this way, Prinzhorn identifies a binding nexus between bodily movement and emotion: "Expressive gestures have the capability of so realizing psychic elements that they are communicated to us directly, as participants. Any motor discharge can be a carrier of expressive processes, not just voluntary movements, but also physiological reflex manifestations such as blushing. From the purposeful movement of the arm, the gesture provoked by joy or anger, to the 'oral gesture' of the word and its manifestation in writing or in a picture, the individual-psychological element is always communicated to us simply and directly, instead of by rational association."[54] But he agrees that the impulse to draw is not in itself an impulse to represent. Children will play with breadcrumbs and mold them into shapes without meaning to re-create anything;[55] they are in fact unaware, until about adolescence, of the representational nature of the artwork.[56] Hungarian photographer

Brassaï clearly grasps this: infantile manipulations manifest a primitive version of his *Involuntary Sculptures*, made of identical breadcrumbs and metro tickets. The visual expression of the motion of the soul is in no way linked to the reproduction of bodies that we encounter in reality. It is a nonrepresentational visualization of the interiority of the human taken as a pathic agent.

Prinzhorn's book drew the interest of Paul Klee, who, in leafing through it, exclaimed: "This picture is a fine Klee."[57] Previously, in 1912, at the exhibition *Der Blaue Reiter* in Monaco, Klee had invited viewers to consider ethnographic collections, clumsy drawings made by children, and creative works by people suffering from mental disorders as phenomena that were helping to reform painting, especially in conjunction with the developments of Kandinsky's art.[58] At the same time, Hal Foster has described the portraiture of the Viennese Secession—Gustav Klimt, Egon Schiele, and Oskar Kokoschka—as a "symptomatic portraiture" in which the figure "becomes a cipher of psychosexual disturbance."[59] Yet this very anatomopathology of the image runs the risk of confusing the symptomatic quality of the artwork with the work itself taken as symptom. On the one hand, Klee's proposal allows us to piece together a story of family resemblances. We might stress, among the many emergences, the first exposition of Art Brut in 1949, in which half the pieces on display were made by psychiatric inpatients. In the exposition's catalogue, Jean Dubuffet wrote that "there is no art of the mentally ill any more than there is an art of dyspepsia or of bad knees." This attitude proved seminal in the context of the 1995 Venice Biennale, which featured Brazilian artist Arthur Bispo Rosário, a paranoid schizophrenic who had been committed to a mental institution in his thirties and lived there for fifty years without taking medications, working on some thousand pieces—including fabrics, banners with geographical maps, spoons, bottles, and brooms—collected in the hospital. This practice would place him somewhere between Arman and Duchamp. On the other hand, it is troubling that Klee's opinion was so closely echoed by Joseph Goebbels, the Reich Minister of Popular Education and Propaganda, who in 1937 exhibited Klee's *The Goldfish* alongside works from Prinzhorn's collection. With Hitler's approval, these were set up for ridicule in the fifth room of the exhibition on *Entartete Kunst* (Degenerate art), which collected works by Ernst Ludwig Kirchner, Marc Chagall, Franz Marc, Dix, Kokoschka, Kandinsky, and Corinth.[60]

Did Goebbels and Klee really see the same paintings? Tobin Siebers writes, in his seminal book on disability aesthetics, that "if modern art has had such an enormous success, it is because of its embrace of disability as a distinct version of the beautiful. The Nazis grasped the nature of this aesthetic, but they rejected it."[61] However, apart from the recent perspective of this field, with its political and theoretical stakes, neuroartcriticism is not about marginalized, impaired bodies in representation, nor does it treat the illness itself as a pictorial theme.

The focus rests on the idea that what counts are the laws of neurobiology. On the healthy side, "art and the production of beauty, in its many guises, is therefore a display of the artist's virtuosity. And it is for this reason . . . that virtually all publicly displayed art throughout human history has been produced by sexually mature males—that is, by men who were energetic and healthy, who had high creativity and hand-eye coordination."[62] Rather than critiquing such a claim with insights from the anthropology of art and from a critical history of the production of beauty, I will go ahead and take it at face value. It is hardly a coincidence that Maffei and Fiorentini cite biologist Desmond Morris's *The Biology of Art* in their work to support the idea that certain aesthetic tendencies reach across the species barrier. Morris indeed finds in other primates behaviors similar to those of humans; this discovery would not have bothered Prinzhorn, who considered biology an invaluable aid in his research. In 1957, Morris, with the backing of art critic Mervyn Levy, whom he had to thank for his own successful career as a painter, held a painting exhibition at the Institute of Contemporary Art in London. The artists on display were Congo, from Britain, and Betsy, from the United States. Both went on to achieve great personal and commercial success—the first for pencil-and-brush drawings, the second for finger paintings. As it happens, Congo and Betsy were both monkeys. The exposition represented an important step in the study of the neurological bases of graphic production (psychiatrists later attributed one of Betsy's paintings to a schizoid ten-year-old girl).[63] According to the biologist—for whom pictorial art, being deprived of its documentary function by photography and freed of all religious purposes, has reverted to a purely expressive value—"the ape and the modern man have much the same interest in producing pictures, and it might even be argued that the modern human artist has little more reason for painting a picture than does a chimpanzee."[64] He goes on to identify six "biological principles of picture-making," which "cover everything and everyone from Leonardo to Congo":[65] self-rewarding activation, whereby the work is a reward in itself; compositional control, whereby uniformity, symmetry, repetition, and rhythm are preferred; calligraphic differentiation, whereby distinct shapes and lines develop by a slow process; thematic variation, whereby a theme is first tried and then worked out in successive variations; optimum heterogeneity, whereby "at a certain point" the painting is finished, complete, and sufficiently colorful; and universal imagery, which hinges on the fact that certain arm movements are satisfying from a motor standpoint while others are not. This last principle, choreographic in nature, seems evidenced by the fact that around 1958, after two years of rich artistic production, Congo's passion for picture making waned: he now preferred to exercise his body and move around. This might lead us to conclude that if we are going to take graphic and pictorial expressions as neurobiological manifestations, independent of stylistic grounds and medial environments, what matters is not the production of a painting but the execution of gestures and movements, including vitalistic ones.

Such is, for some, the principle behind Jackson Pollock's "action painting." When Maffei writes that the artist "pours out on the canvas unconscious mental images, nested in his motor nerve structures, which are expressed in his dance around the painting, in a motor fury that becomes art in the form of signs and colors" and that his paintings "*are* the reality of our motor memory, they are the familiar gestures of dance turned into images," he aligns the poietic power of a work, the efficacious skill of its maker, and the effectivity of beholding with the fact that this experience is possible precisely because Pollock did not dance *in order to* transfer his mnestic-motor rhythm into his paintings, but rather embodied the brush and made himself into an instrument of painting, as has been apparent ever since Hans Namuth's 1951 film about him.[66] To point out that Salvador Dalí already saw Pollock's gestures as monkey-like is to mistake a taunt for an explanation. Pollock did not wave his arms over the canvas, nor did he pogo aimlessly; he went in and out and stopped when the work's completeness demanded it. His drippings are the expression of a formal, formative finality. I therefore believe that neurology's attempt to study us by way of the material culture we produce, while simultaneously disregarding this same materiality, begets a sterile paradox whereby the "neuro-knowledge" produced winks at an oddly literal, nonreflective knowledge.

## CEREBRAL EVIDENCE

Not only does the survey of the specificities in the pictorial production of brain-damaged artists set up a *Wunderkammer* of cerebral deficiencies, but it also suggests an entire systematics to be built around medical accounts, based on the type of brain lesion, its localization, and the ultimate output, with no consideration for artworks' media, materials, and histories. From the standpoint of beholding, artworks are considered "stimuli intentionally designed to trigger a range of affective, perceptual, and semantic responses in consumers constitutive of their artistically salient expressive, formal-compositional, aesthetic, and cognitive content,"[67] but from the standpoint of neuroartcriticism, once it is assumed that neurological deficits are special abilities, the work itself is treated as neurobiological evidence of the artist's cerebral conditions, even where no reliable account is available. In Chatterjee's words, "art is worth considering as a neuropsychological probe."[68] Artworks become material for a kind of neurological autopsy of their makers (as representative of humans). Their poietic nature, though nominally accounted for, is fundamentally misconstrued. This brings to light an astonishing similarity to the psychoanalytic interpretation of art, to which Gombrich, in 1953, after acknowledging the undeniably personal nature of any single artistic production, addressed the following objection: "But does it matter all that much? . . . For try as we may, we historians just cannot raise the dead and put them on your couch. . . . We historians could always prove to you that the information you need is not to be had, and you could retort that without such essential information we might just as well pack up and go home. And so I repeat the question whether it really matters all that much if we know what the work of art meant to the artist. It clearly matters on one assumption and on one assumption only: that this private, personal, psychological meaning of the picture is alone the real, the true meaning."[69] His authoritative statement notwithstanding, neuroartcriticism, too, looks beyond the artwork to grasp the unconscious mind of the artist; it thus proves indifferent to the artifactuality of the work, which is treated as if it were an encephalogram or a scan, a proxy.

We observe a lone woman with milky skin, wrapped in a white robe, sleeping supine on a bed, her neck and blond head falling over at the right of the painting, her arms draped back along the sides of her face.

A hideous little brown monster, difficult to identify, crouches on her stomach with its feet pressed against her sternum; we see it from the side, but with its nasty face turned toward us. There is a horse's muzzle protruding from behind a red curtain at the back left of the room. *The Nightmare* (1781) by Heinrich Füssli reproduces the iconography of sleep paralysis coupled with hypnagogic hallucinations, which affect the senses of sight, sound, and touch. Hobson and Wohl, while careful to point out that in the late-eighteenth-century Swiss painter's homeland nightmares were still commonly believed to take the form of a horse staring at the sleeper, and that the incubus—the little monster on the sternum—has served as an icon of the nightmare since antiquity, argue that the two beasts are "[products] of the periodic internal activation of the brain and the cardiocirculatory system," despite appearing in the painting as "mere symbols of these internal brain processes."[70] The specific difference between pictorial figure and hallucinatory image is thus accepted and recognized, but ultimately remains little more than a concession: "OK, maybe neurons and pictures are not the same thing, but still . . .". In a similar vein, Frank Clifford Rose argues that the sharp-looking boy in José de Ribera's *The Clubfoot* (1642), shown from a three-quarters angle with a paper in his left hand that reads "Give me alms, for the love of God," suggests infantile hemiparesis. The painting might even be the first portrait of an aphasiac, given the written message. However, the British neurologist concedes that if the right hemiparesis had been caused by an injury to the left lobe during childhood, the right hemisphere would have taken over the linguistic abilities of the left, rendering the message unnecessary. He thus concludes that "either the right hemisphere was also affected or the note may be a ploy to obtain sympathy and 'alms.'"[71]

This second strand of neuroartcriticism ends up treating artistic production as direct and inescapable evidence of a cerebral pathology. But instead of taking up works to study the representation of disease symptoms within the history of figuration, it directs its focus to the nexus of art production and cerebral malfunction, almost as an unconscious expression—a kind of lapsus—of the affinity between the malleability of the material and the plasticity of brain. To be sure, researchers acknowledge that "in some cases, the depiction of an apparent disorder may merely reflect the artist's intention to convey an emotion or idea rather than a specific medical condition."[72] In this vein, Chagall portrayed himself as a polydactyl on his left hand in *Self-Portrait with Seven Fingers* (1912–1913), despite having the usual five fingers. Nevertheless, at the heart of such a reading, we encounter the representational magic whereby the painted bodies themselves are sick. The possible documentary value of the painting, as legitimately investigated, gets confused with its own exceptional character as an artwork, which, conversely, we grant only because the painting in question is already a part of the artistic tradition. For its part, the work, in all its singularity, becomes a mere channel for expressing a particular content: the sufferer's disease. Neurobiological malfunction as the impetus for an artistic genre has even played a role in the selection of curatorial themes, thereby reinforcing the idea of a systematics of art based on neural merits. The website for the Deutsches Epilepsiemuseum hosted by the Handwerksmuseum in Kork, near Karlsruhe, groups together paintings that represent epileptic seizures. These include, among others, Raphael's *Transfiguration*, the primary iconographic reference point of this successful pictorial genre, whose figures are no longer thought to show suffering, visionary, or ecstatic men, as Nietzsche argued they did,[73] but rather to constitute material evidence of hysterical poses. There seems to have been no shortage of attempts to grant equal relevance to the representation of fainting in art.[74] We should bear in mind that a criterion of this nature is both applied to various works and seized upon by some contemporary authors as a principle of recognition. This is the case for Jennifer Hall, who collected the outputs of twenty-seven artists with temporal lobe epilepsy; *From the Storm: Artists with Temporal Lobe Epilepsy* opened in Boston in 1992 at Do While Studio and was subsequently exhibited during the International Epilepsy Congress in Sydney in 1995. The disease itself drives the making. Another genre is migraine art. It sprang up in the 1980s in England when the British Migraine Association and the pharmaceutical company Boehringer Ingelheim sponsored a painting competition calling on migraine sufferers to submit realistic representations—as opposed to artistic renderings—of their attacks. Nine hundred paintings streamed in from all over the world, and the collection currently houses 562 of them. The concept of migraine art had been the brainchild of Derek Robinson, who argued that the techniques of pictorial representation provide an apt tool for expressing and communicating symptoms of the disorder and the way sufferers respond to their pain.[75] Klaus Podoll, one of the leading researchers of the genre, established the prevailingly diagnostic value of this kind of work in a study of 450 migraine-suffering artists, all the while claiming that the impact of this condition on the history of art has been comparable to that of dreams and drugs.[76] Dubuffet would be appalled!

But if we reverse our viewpoint and turn to the image as such, the question becomes about whether there is a specific causal correlation between neuropathology and iconography, irrespective of the image's therapeutic utility or expressive practices. Once again it was Podoll who, in a study on Giorgio De Chirico carried out with Ubaldo Nicola, made a positive attempt to respond; the authors sought to establish a "neurology of creativity" that did not rule out the possibility of other, sociohistorical contributory factors.[77] De Chirico suffered severe headaches with extravisual hallucinatory auras and, despite lacking a formal medical diagnosis, described its symptoms in detail. Among its characteristic features was the presence of scotomas—a kind of "black hole" in the visual field. Podoll and Nicola's hypothesis is that the emphasis on shadows, found in works such as *The Red Tower* (1913) and *The Departure of the Poet* (1914), results from the painter's direct experience of this visual phenomenon. The same is said to apply to the auratic experience of time, which is at once dilated and punctual, however dense and stable it may appear in *The Enigma of the Hour* (1911). The authors further theorize that *Self-Portrait* (1920) represents the experience of a parasomatic body: I am still I, but at the same time my body is outside my skin and a different size; it has, as it were, a degree of autonomy. People suffering from auratic hallucinations know that their content is not real, yet the effectiveness of the perception is overpowering; in addition, the auratic experience enlarges the sphere of the neurobiological dimension of the Self, even though the person retains a full and lucid awareness. To support their interpretive hypothesis, Podoll and Nicola turn to the painter's writings. In one instance, De Chirico asks the reader to imagine entering a room and seeing it with all its furniture properly arranged: a man sits in a chair; a cage hangs from the ceiling; paintings are mounted on the wall. Suddenly and involuntarily, the relationship between the things breaks down. Its nexus falls apart: "that is not to say that the scene has changed, but that I see it from another angle." For the Italian painter, everything has two aspects: an everyday aspect and one that is "spectral and metaphysical," akin to that of "certain bodies hidden by matter impenetrable to sunlight, which only appear under artificial light such as X-rays."[78] But contrary to what the authors suggest, the reference to X-rays does not constitute evidence that a migraine-related disconnection between memory and perception provides direct access to objects, as though De Chirico saw them for the first time and therefore painted them as at once strange and familiar. Even if the auratic experience effectively creates the sense that objects are in relief with respect to the surface of the visual scene, this does not imply a *direct* access to the artistic image *through* the clinical biography of its maker.

This overlooking of the image's artifactuality is a function of brainocentrism and the related idea that the brain is the actual master and maker of our life, but it is not per se a necessary outcome of the medical knowledge deployed in the inquiry. The art-historical method of Giovanni Morelli suffices to prove that a diagnostic approach is not fundamentally reductionist. Morelli, in addition to being a great admirer of Goethe, held a medical degree with a specialization in comparative anatomy. In order to correctly attribute artworks to their makers, he devised a working method based on what Carlo Ginzburg later called a "circumstantial paradigm." This model is viable for the humanities insofar as it helps resolve the sharp opposition between an inquiry based on a "grounded" quantitative rationalism and one based on an "aleatory" qualitative irrationalism. The maker's signature, Morelli concluded, was discernible from details of the body that fall outside the conscious compositional will of the creator—for instance, in hands, ears, and halos. This led him to look at the body's painted shape rather than at its structure. His ultimate goal was to interpret the morphological and functional significance of stylistic forms, such as the nails that do not fully cover the flesh on the fingertips of the Madonnas painted by Raphael, whose work may also be recognized by the figures' round and fleshy ears.[79] To do so, he would compare each ear in a series of such features, and—by opting to examine its phenomenology, its appearance, or its modes of appearing—succeed in verifying its formal autonomy. He thus guaranteed the authenticity of the work by really looking at it.

While Morelli aimed at the correct attribution of works—and many of them were made thanks to him—Freud, declaring his own interest in the Morellian method in "The Moses of Michelangelo" (1914), makes use of the same principles to discern the "artist's intention" (though not his private state of consciousness). His study of Michelangelo's statue begins by considering the position of the right hand—its index finger alone holds back Moses's beautiful bushy beard, gathering up its left part—and the horn edge of the tables of the law clasped against his side by his hand and arm. From these clues, the psychoanalyst reconstructs the story of a movement involving the entire body, along with the mental impulse that produced it. What we are made to see, in the end, are "traces of suppressed movement."[80] Freud concludes by inferring the reason behind Michelangelo's compositional choice: the artist, being impetuous like Pope Julius II, for whose funerary monument the work was commissioned, was reproaching his deceased patron for his wrath while also admonishing himself. Freud, while doubtless wanting to understand the maker's personality, does not forget that his analysis is that of a beholder using his own imagination to complete the morphological clues—the figural symptoms left by material and perceptual traces, such as the unusual position of the index finger. He thus does not lose sight of the artifactual quality of the work itself. At last, a difference emerges between a psychoanalytic approach to art and neuroartcritical naturalization.

Studies of neuroartcriticism often contain disclaimers stating that while there are doubtless many epileptics and migraine sufferers, not all are artists. This acknowledgment may seem like a gesture of resignation in the face of the inexplicable. Yet the real source of discouragement is an approach that looks at art through a naturalistic lens, when the work as such is reality rendered in an artificial way, not its literal confirmation. As Marc wrote about his bronze sculpture *Two Horses* (1908–1909): "I have intentionally tried to remove any indication as to type and breed from these horses. Therefore, for example, the vigorous proportions of the limbs are somewhat unhorselike."[81] This is not paradoxical: the work is the result of a principle of selection and abstraction applied to figurative schemata with formal properties, which indeed induce pathic effects but are not in themselves sorrowful or painful. This is why the text *Laocoön*, in which Lessing examines the pained expression of the priest crushed by snakes, is a study on the limits of pathographic expression in art and not on the causal power of human suffering. In other words, it seems to me that pathology, in an informed neuroartcriticism, should serve as a barometer for studying the logic of the pathos, the hurt, the pain that the work controls, restrains, and thus engineers. Otherwise, we neglect the fact that, while there are many, varied reasons for which one might represent one's condition, in the case of an artwork, this representation is bound to the world by the formal freedom the artist has conferred on it by creating it.

This reasoning recalls the dualism of nature versus culture, which neuroaesthetologists largely seek to dissolve in favor of a monistic naturalism. Often cited in support of this position is John Constable's fourth lecture, given at the Royal Institution on June 16, 1836. The painter takes up English portraitist Thomas Lawrence's claim that we cannot hope to compete with nature when it comes to the beauty and elegance of its forms and colors, adding: "Painting is a science and should be pursued as an inquiry into the laws of nature. Why, then, may not landscape painting be considered as a branch of natural philosophy, of which paintings are but the experiments?"[82] If such is the case, then Kandinsky's *Cossacks* really should be thought of as a perceptual experiment, or a Rorschach test. Constable, however, studied Alexander Cozens's drawing manual, *A New Method of Assisting the Invention in Drawing Original Compositions of Landscape* (1785), and put its schemata into practice drawing clouds. It is true that Cozens proposes criteria akin to those formulated by Hermann Rorschach in the methodological introduction to his *Psychodiagnostics: A Diagnostic Test Based on Perception* (1921), calling for simple and symmetrical images to be drawn up and distributed "rhythmically" over the paper surface, which the subject—who is asked, "What might this be?"—can turn over at will, holding it at different distances from the eye.[83] But the Swiss psychiatrist emphasizes that the subjects—particularly neuropaths—do not interpret the images but rather describe them, set them up perceptually, and only afterward potentially transform them. Conversely, when Cozens invites us to start from an indistinct blot that has been deliberately thrown onto the page and allow the contours of an idea to "suggest" itself, he is conceiving of the final image as the perceptible figure of a visual pattern, as the perceptual material from which a clear form may be artfully extracted. In his *Treatise on Painting*, Leonardo calls on the painter to discern wondrous stimuli for clever inventions in mold stains on the wall, in ashes, in clouds, and in mud, but criticizes Botticelli for abusing the practice. As he points out, the eye will find, in a colored stain on the wall, whatever it is looking for, primarily landscapes; yet little can be gained without a study of the particular, and "Botticella" indeed painted "sorry [landscapes]."[84] The pictorial spot is not visible reality, flattened in its perceptibility. It suffices here to recall Gombrich's sensible observation that what Constable learned from Cozens was not how clouds really appear "but a series of possibilities, of schemata, which would increase his awareness through visual classification."[85]

The artwork is not a natural fact once and for all, and the access it provides to the world is not direct; if anything, it is sensible. In this way, the disease's artistic value lies in its capacity to elicit a nonliteral understanding of reality. When Podoll and Nicola, in studying migraines, note the sense of verisimilitude, of perceptual accuracy, presented by De Chirico's metaphysical and auratic paintings,[86] they reveal their sensibility as beholders to the performativity of the work. Not coincidentally, Waldemar Conrad, the first aesthetologist of the phenomenological school, who in *The Aesthetic Object* (1908) sets out the ideal and formal properties of the artwork, observes that it is legitimate to derive compositional principles from the way the eye behaves. Aesthetics, he points out, starts down "the same paths as the natural sciences," which begin from the "point of view of ordinary life," but is soon forced to diverge because it is driven by a completely different principle of causal connection, lest it be compelled "to substitute complexes of atoms for sculpture and air vibrations for music."[87] The diagnostic focus on the artwork as an impetus for neurology gives rise to misconceptions because the type of analysis undertaken by neuroartcriticism is devoid of iconographic training and is really an analysis of the making and beholding process. It is unconcerned with the artifactual quality of the product as created and looked at. It therefore boils down the final work to a mere careless expression of human neurobiology, akin to a lip print left on the rim of a glass. Moreover, while the degree of causal necessity that links neurological disorder and image making is clearly one of the stakes of studies in neuroartcriticism, these studies, when taking care to avoid being *crudely* reductionist, paradoxically end up not being *rigorously* reductivist. In other words, they do not really attempt to derive culture from nature, as they supposedly set out to do, and as Du Bos and Taine did to some extent in previous centuries with their climatic theories, which we would today call environmental aesthetics. Neurological studies of the artistic process remain tied to the idiosyncrasy of the single work, the single datum, leading even Chatterjee to concede that "the neuropsychology of art has not developed into an experimental science" and that it needs to go beyond the mere cataloguing of anecdotes, however fascinating they may be: "These rich qualitative accounts need quantitative underpinnings."[88] But for this to happen, it would be useful to clarify the sense in which the human neurobiological organization can potentially be normative, and how an explicit deviation from the regular equilibrium could allow us to understand such a normativity.

In its present state, neuroartcriticism attests to the scope of the brain/art parallelism at play in neuroaesthetics, whereby "visual art can be decomposed into distinct attributes such as color, line, texture, and form,"[89] but it does not serve the purpose of "recognizing the ontology of the accident," which is a difficult, philosophical task, because "it must be acknowledged as a law that is simultaneously logical and biological, but a law that does not allow us to anticipate its instances."[90] The search for the rationale behind the variety of human production continues.

# 5

# THE BRAIN'S ICONOCLASH

Neuroaesthetics would not be possible without the technologies that make visible our neural processes and the iconic effectiveness of the images they generate. Any further inquiry into its epistemic status and heuristic capacity thus ought to be complemented by a discussion of the visual media used to display cerebral states and activity. This is achievable by interrogating the neurological and medial routes whereby mental images become material images and, conversely, images from the external world are elaborated cerebrally. The mutual collaboration between art production and brain-imaging experiments during the first two decades of this century have brought these processes increasingly into focus. As I mentioned in posing the question of the cultural value of neuroaesthetics, in 2003 the conceptual artist Jonathon Keats collaborated with neurologist Bruce Miller to image his brain while he contemplated beauty and truth. Fourteen years later, in 2017, neuroscientists Yukiyasu Kamitani and Tomoyasu Horikawa developed an fMRI method to show how brains that have seen or imagined pictures produce signals that can be reverted into digital images corresponding to those seen or imagined. As they demonstrated, "visual features extracted by computational models were successfully predicted from brain activity patterns."[1] The circularity between cerebral activity and pictures becomes apparent to the external observer thanks to visual patterns whose production depends on multiple makers, not all of which are human, since a wide range of tools and computer applications also play a part. These consist of complex computational systems involving physics, chemistry, biology, computer science, electrical engineering, neuroanatomy, and other complementary areas of knowledge. As the expression of this visual fabrication, the neuroaesthetic brain is thus confirmed to be a *living* (i.e., organic and temporal), *discursive* (i.e., epistemic and political, disciplinary and institutional), and *medial* (i.e., materially fabricated and rooted in one or more apparatuses) artifact. Although its specific construed nature defies scientific representations from previous centuries, it shares a decisive feature with them: they are all "produced as 'similitudes' to other . . . representations"[2] and do not derive their reality effect from any possible resemblance to the natural model. That notwithstanding, our brain is depicted and held up as the ultimate icon of scientific objectivity, as its visual validation.

Brain images arouse both admiration and hatred, as icons have always done, and this engenders two interconnected clashes. The first concerns images' capacity to classify mental processes as human—putting a face on them, as it were; the second, their capacity to bring them close to us, as near to hand as a portrait on the night table. These clashes are experienced not just by researchers in the neurofields but by lay onlookers as well. On the one hand, ever since "the brain [started being] considered [a] *locus* of subjectivity,"[3] brainhood itself has acquired an indexical value. Thus, when philosophers such as Patricia Churchland say, "I am who I am because my brain is what it is,"[4] the statement could actually be reformulated as "I am my cerebral area whose activation mirrors my suffering." This is the spirit of Daniel G. Amen's clinic, where neuropsychiatrists show Single-Photon Emission Computed Tomography (SPECT) images of the brain to suffering patients to explain to them what their painful personalities look like.[5] On the other hand, brain images pit technicians and researchers, who advocate a quantitative definition of brain imaging—notably, of functional imaging[6]—based on strings of numbers and diagrams, against clinicians, who rather endorse a pictorial understanding. Scholars in the first line "are not entirely iconoclastic in that they do accept that images play a role, but they do fear too great a reliance on the icon—a seduction of the senses and the neglect of the greater truth that lies beyond representations,"[7] from neuroscientific and technological standpoints as well. In any case, even though "for brain mappers, the visual is a nonstarter,"[8] since scanners are not an optical technology, neuroimaging produces actual pictures, which have the effect of placing the brain right in front of us. They make it graspable. Finally, where neuroaesthetics is concerned, the never-ending confrontation between iconophiles and iconoclasts, once waged around the most sacred religious images, plays out on the profane ground of the present-day neuroscientistic faith between those who think of neuroimages as snapshots of a person's thoughts and those who object to the visual practices implied by such a reductionist conception, thereby conceding that mental activity can in principle be reduced to its portrayal—no matter how misleading it is—and that brain imagers act like Trojan horses in our personal fortress. One way or another, we trust images, and this trust is based on the pictorial understanding we have of them. Here, I intend to discuss the conundrum whereby such a reliance is disrupted by the rise of a new scopic regime made possible by brain imagers, whose visual outputs owe their rhetorical efficacy to the very iconic habit they disturb. These conceal themselves beneath the familiar appearance of the brain matter we take from corpses, when really they produce an unprecedented visual artifact.

To do so will require suspending widely held assumptions, canonized by Merleau-Ponty, about a pure scientistic attitude that takes the researcher's gaze to be neutral, objective, and natural. Such an approach is by no means mandatory, despite what Churchland and similar thinkers would have us believe. I am inclined, rather, to reframe the ideological issue as a visual and a medial one. As Jonathan Crary taught us well before neuroaesthetics emerged on the scene, "Virtual environment helmets, magnetic resonance imaging, and multispectral sensors are only a few of the techniques that are relocating vision to a plane severed from a human observer."[9] There seems to be a dehumanized, disembodied, and yet material gaze at work: are we once again confronted with the narcissistic display of old, whereby onlookers gazing at themselves neglect the water's surface, failing to acknowledge the artifactuality of the visual outcome, or have we merely entered a highly specific phase of the post-photographic era—namely, a sort of cyborg scopic regime that ought to be the very focus of our critical analysis? To answer this question, we need to interrogate more thoroughly the claim that the neuroaesthetic brain is a by-product of neuroimaging. I am not saying simply that cerebral images are "operational"—that is, that human sight is obsolete and imagers see autonomously, as in filmmaker Harun Farocki's already well-established notion. Nor am I just stating that the ensemble of material images commonly addressed by visual culture cannot ignore neuroimaging (for indeed, the very notion of the brain image as an image per se is now being questioned by neuroscientists such as Damasio in debates within visual culture).[10] Rather, I am persuaded, with Bruno Latour, that in visual production, "the more mediation, the *better* the grasp of reality."[11] As I will argue, neuroimaging is not a matter of the straightforward visualization of embrained processes, and brain images are not the *longa manus* of naturalism but the actual production of a visuality that forges our lifeworld. Media matter; and if media matter, there can be no brain matter without visual media.

## CYBORG VISUALITY

Ever since the human brain joined the debate about Selfhood, the question has crept in whether "[the time is] right for the neuron to step up and take its place in twenty-first century visual culture."[12] If the editions of the Brain Art Competition,[13] held since 2011, offer any indication, the answer is undoubtedly yes. But a major sticking point arises from the fact that neurons are invisible to the naked eye and to an ordinary microscope, despite the progressive extension of the figurable achieved over the centuries. To be sure, research in the life sciences has always needed to make its objects of study available to sight. The requirement of having effective solutions for visualizing unvisible things—things that outstrip human vision because they exceed our visual field or are not visible enough—is widely averred by scientists, who risk feeling lost and powerless otherwise. What is most compelling, though, is that *unvisual* bodies have been made visible. Emblematic is the case of quantum mechanics, where Niels Bohr tells of the constraints arising from the natural limitations on our forms of perception, which are restricted spatiotemporally and with respect to their own acuity[14]—a complaint reminiscent of Professor Kevin Warwick's, as mentioned in chapter 2. Yet Karen Barad questions the Danish physicist's reasoning, notably his conception of apparatuses as "macroscopic material arrangements through which . . . particular phenomena with particular determinate physical properties are produced";[15] she objects that "Bohr mistakes the apparatus[es] for mere laboratory setup[s]"[16] and promotes her own understanding of the same as medial, material, and discursive practices that "enact what matters and what is excluded from mattering."[17] Every apparatus (re)produces specific boundaries and assigns properties to phenomena.

In this line of reasoning, the traditional meaning of visualization gradually changes in accordance with the technologies employed. It is no longer about placing under a better light, putting in the proper perspective and at the proper scale, and so on, but about making visual. It is about *visibilization.* Such an evolution has been neatly thematized by the work of Felice Frankel. Let us consider, from the collection of images developed by the MIT biologist, *Ferro Fluid*, a suspension of magnetite in oil that, when placed on a yellow-tinted glass resting on seven magnets that attract the fluid, looks like a raspberry on a layer of cream; *Yeast Flower*, a yeast colony masquerading as a white camelia; and *Adhesion*, where the tearing of a ribbon from an adhesive surface creates ornamental and zigzag patterns. In the introduction to her Italian catalogue *L'incanto della scienza* (The enchantment of science) (2004), she writes: "Like the equations in mathematics and the structural formulas in chemistry, these photographs are meant to communicate ideas. They are a visual representation of physical phenomena that reveal hours of deliberation by scientists, years of preparation, and a lifetime of discovery."[18] Communicating ideas entails suggesting schemata of possible structural arrangements in organic matter and eliciting similarities between shapes that recur across unrelated materials. However unwitting the link, Frankel's work follows in the spirit of Karl Blossfeldt, a teacher of live plant modeling, who, between 1928 and 1932, published black-and-white blowups of selected, stylized leaves and stems, whose forms had been simplified and subjected to the metamorphic effects of the photographic lens-eye, which could magnify and reduce. We can thus make out the tip of the Chrysler Building in an *Equisetum majus* (the Empire State Building more closely resembles the *Equisetum hyemale*), while the *Adiantum pedatum* looks like an Egyptian *Heq-At* (hooked scepter) or a crosier. Microscopic forms of the marine world were the reference point for Ernst Haeckel's color lithographs, produced between 1899 and 1904, on which Klimt and Klee trained, and whose jellyfish, coral, algae, and starfish captivated the exponents of Art Nouveau.[19] These are all images that render the natural world without any mimetic aim.

But if the proposal to re-create the world by looking at it is not new, our mastery over the visual is. Techniques that allow for the visibilization of blind spots serve as spatiotemporal prostheses of our senses, even opening us up to unvisual worlds. Electronic images—of the nanoworld, for instance—are objectless images. Whether colorized or not (they are originally colorless and receive their saturated and flashy hue from commercial software), they are "separated from their scientific source, in several ways," as Roald Hoffmann, Nobel laureate in chemistry from 1981, notes; "they depict the very small—500 nanometers (nm) across is the typical size [whereas a] baby's hair might be 25,000 nanometers thick."[20] Certain images of the nanoworld are seen not by light but by means of a probe that responds to the magnetic properties of matter. These, too, are images of objects that have been interfered with, either through a reforming of their material by technology, as in the silicon nanolever, or via processes of self-organization, as in the bundling of carbon atoms. What happens, then, when the unvisual is the property not of matter but of (cerebral) activity in the process of becoming, so that the human observer and the one observed come face to face?

The mirroring technique represents the first chapter in the media history of self-gazing. Plautus, in his *Epidicus* from the second century BCE, states that an old man sees his past mistakes reflected in the mirror.[21] According to Diogenes, who wrote some five centuries later, Socrates used a mirror to show drunkards their faces disfigured by wine.[22] In those days, the mirror was considered a rare and precious thing, a device for self-knowledge, purportedly invented by Hephaestus, the Greek god of fire and metal. For much of its history, from the manufacture of pocket mirrors to the installation of reflecting surfaces in public spaces, the mirror has been thought of as a medium for externalizing a person's inner life.[23] Without its help, we would have remained blind to ourselves. Now, however, as our contemplation shifts from reflecting an individual interiority to visibilizing cerebral processes, the personal loop of self-inspection becomes an open exhibition. I intend to interpret the evolution just charted in terms of the technical improvement of more basic devices. I reject the idea that medical imaging techniques "belong . . . to a wider history of tools for seeing and visualizing, including mirrors, the camera obscura, photography, film and video cameras,"[24] since that would be true only in a very general sense. Rather, they make visible by forging vision itself. Brain imaging is not a perfected visual strategy but an image-making practice, technologically heterogenous to previous visual media based on a representational principle.

For centuries, the brain, which we now take to be the organic basis of our mind, was detectable only thanks to postmortem autoptic examinations and the rough renderings to which they gave rise. It must be remembered that "ancient investigators had no effective way of visualising the [ventricles'] complex shape after the brain was sliced open. . . . Simple knife cuts only provided a minimal amount of information concerning their layout, a situation that encouraged medieval illustrators to portray the ventricles as simple cells with no effort at anatomical accuracy."[25] The psychologist Andrew P. Wickens singles out "the emergence of intaglio technology, most notably woodcut illustrations and printing in the fifteenth century" as the most important breakthrough for the study of cerebral functioning because "with this development we begin to see the publication of the first dissection manuals, books on anatomy and depiction of the brain."[26] This innovation culminated in the 1543 work *De humani corporis fabrica* by Andreas Vesalius, in which the Flemish anatomist described, to a certain degree of precision, the brain's morphology, which he accounted for with anatomical tables painted by him and other contemporary masters. The complexification of the medial process has increased with advances in visual technologies. In 1685, Malpighi boiled the brain and, after removing the meninges that separated the encephalon from the braincase, soaked it in ink to produce a contrasting effect between its parts. A century and a half later, discoveries were occurring at great speed. The physiologist Franz Joseph Gall is best known for his *Influence of the Brain on the Form of the Head* from about 1825; having observed large, protruding eyes in many of his acquaintances with good memories, he posited a link between the bulges in the skull and the corresponding cortical functions, and thus invented cranioscopy. The localizing principle Gall used was both fantastical and flawed, except with respect to the language area, which was not too far from the one identified today. In 1894, a New York doctor, J. W. Redfield, launched a similar project, publishing a diagram of the human face and brain that listed 160 faculties, the 149th being "republicanism." The artist Lara Favaretto ironically reprises this idea in *Thinking Head* (2019) by localizing functions such as animism and self-doubt on the skull.[27]

What is often overlooked, however, is that one of Gall's main innovations "was a novel form of dissection," in which "instead of slicing the brain in sections as most others had done with a knife," he and physician Johann Gaspar Spurzheim "hardened it in alcohol, and then took it apart from the bottom upwards, teasing away its structures with their fingers."[28] In 1854, the anatomist Joseph von Gerlach started using carmine, a pigment derived from cochineal, to stain cerebral tissue, which turned red if treated with ammonium carminate. The stain produced a clear differentiation of nerve fibers and cells. As Wickens observes, "[von Gerlach's] discovery is generally recognised as representing the modern founding of neurological staining."[29] These discoveries notwithstanding, "the structure of the brain resisted visualisation" until 1873, when cytologist Camillo Golgi accidentally spilled a silver nitrate solution on cerebral tissue previously hardened with potassium dichromate. Doing so caused the material to darken and the nerve components to stand out against a silvery-yellow background. This gave rise to a new staining method, the so-called black reaction, which colored the cell down to its tiniest branches of dendrites and axons and made it possible to highlight a few individual cells in any given piece of brain tissue. Whereas in the early days the process proved highly time-consuming and capricious, it was subsequently improved by Santiago Ramón y Cajal—an expert in color photography, an aspiring painter, and the source of some three thousand drawings of the microscopic anatomy of the brain—who also developed several techniques, the most notable being the reduced silver nitrate method from 1887.[30] Ramón y Cajal's work later caught the attention of the Surrealists via Dalí. In the same decade, the Italian physiologist Angelo Mosso, a close friend of Etienne-Jules Marey[31]—who in 1865 invented the pneumatic pneumograph, which Mosso later modified—established the conceptual basis for noninvasive functional neuroimaging techniques. In 1881, he developed the plethysmograph, a device that could measure cerebral blood-flow variations by recording brain pulsations in patients with skull defects, and concluded that alterations in blood flow to the brain were determined by functional changes. Despite the enthusiastic reception by journalists of his day, as well as by William James, who gleaned from him the idea that an "afflux of blood" occurs in the brain during mental activity, Mosso remained in relative obscurity until recently.[32] Nonetheless, his device can be considered the first noninvasive functional neuroimager. In 1884, Mosso also built a "human circulation balance" to measure cerebral activity during emotional tasks. The instrument, consisting of a wooden plank set over a fulcrum and a system of counterweights, required subjects to lie down, immobile, until equilibrium was achieved; then, as the brain processed a stimulus (e.g., a ringing bell), variations in blood flow could be detected through shifts in the table's balance. In the mid-nineteenth century, Søren Kierkegaard made the following telling reflection: "Soon . . . someone will invent an instrument for listening to the pulses of the brain. That will make a tremendous stir, until in fifty years' time every barber can do it. Then, when one has had a haircut and shave and been stethoscoped (for by now it will be quite common), the barber will ask, 'Perhaps, sir, you would like me to listen to your brain-pulses?' "[33] While it took slightly longer than the Danish philosopher envisioned, such capabilities have indeed become commonplace; now, an entire media archeology on tools and materials could be written, stretching from when Leonardo held the wax cast of the cerebral ventricles to the first steps in the visualization of the living brain in 1918, when Walter Dandy injected air in the place of cerebrospinal fluid, to 1927, when Egas Moniz injected radioactive dye into the carotid artery to sharpen the quality of cerebral X-rays. Over this trajectory, we have gone from visualizing the organ as extracted from a dead body to making embodied neural processes visual: we have gone from corpse dissection to in vivo inspection and from visualizing to visibilizing.

Of course, from depictions in woodcuts to present-day digitalization, different understandings of the brain have arisen. Among the more significant is that portrayed in Hans von Gersdorff's 1517 surgery manual, attesting to a brain dissection that took place in Strasbourg that year: "Individuals were not as much interested in knowing the body as in being seen as they participated in the demonstration of the divine truth of the body," previously learned from books.[34] The matter handled by surgeons and the "true" knowledge about it were assumed to belong to different knowledge discourses. In our time, with the passage from a constructed external gaze to a self-gaze that negotiates between self-awareness and shared biological grounding, a specific conceptualization of the brain's status has also arisen. The tension between the first- and third-person viewpoints is decisive for brain images; after all, cerebral matter is a common denominator of human beings. Indeed, it is by no means trivial to point out that, to be identified, the renderings must be paired with a label bearing the name of the brain's owner. On the one hand, then, images record and classify a given living brain; on the other, they evacuate it from the organism that embodies it. Harvard Medical School offers a free online MRI atlas that simultaneously shows the horizontal, sagittal, and coronal planes of the hemispheres, the brain stem, and part of the spinal cord. In each plane, the user can navigate across the various structural levels, viewed as both MRI (T1 or T2 weighted) and PET images.[35] These renderings are both unattributable and standard. Indeed, many databases confirm the "continuous effort by the medical community to discipline [brain] images and through them the human body."[36] The tension between individual brains and normalized cerebral layouts replicates the stalemate mentioned earlier, with its resulting iconic clashes, whereby brain images are seen as a kind of hyper-objective neuro-snapshots. This was the conflict driving Jean Baudrillard's polemic in his *America* (1986), which targets the "superstition" that "all that fascinates us is the spectacle of the brain and its workings" and that "what we are wanting here is to see our thoughts unfolding before us."[37] Such was certainly the desire of philosopher Herbert Feigl. In 1958, he envisioned an autocerebroscope that would allow people to examine their own brains; at least from the standpoint of thought experiments, this marked the birth of neurophilosophy.

However, beyond the advances in visualization techniques that turned speculative exercises into current practices, brain imaging produces not images but statistical maps. Brain imaging involves not *seeing* but measuring. Therefore, as we go to survey the relevant technologies, we should bear three points in mind. First, the images obtained bear absolutely no inherent relation of optical resemblance to what they show; they are produced by recording systems that report an unvisual reality, but they picture spatiotemporal distributions—for instance, of oxygen—without necessarily being isomorphically related to either oxygen or the brain. Second, neuroimaging techniques render the existence of mental states as an occurrence of neuronal activity, but no commonly used functional imaging method can measure neuronal activity *directly*:[38] there is no longer a subject who poses but a technique that poses its object. Third, the brain's anatomy, functioning, and activity are recorded using procedures that select certain aspects over others, and "an uncritical use of new imaging technology may open the door to a new kind of old fashioned phrenology . . . looking at specific areas only and neglecting the interconnectivity of a neuronal network."[39] The following is a list of the imaging techniques available today. In considering them, we should bear in mind that when the person under inspection lies down inside a scanner, he or she is inaccessible except via spoken instructions, and that the tasks used to elicit specific patterns of brain activity are often experimental proxies for far more complex real-life behaviors. The International Affective Picture System, for example, provides a set of normative emotional stimuli for experimental investigations of emotion and attention. These offer a large, standardized, emotionally evocative, and internationally accessible database of photographs comprising content from a wide range of semantic categories.[40] Moreover, in the visual result, the choice to indicate the site of increased metabolic activity using vivid colors is hardly neutral. I understand it as serving both to draw our attention to the activated area and to convey a sense of intellectual excitement about its recognition.

Brain-imaging techniques can be classified as either structural or functional.

Structural techniques visibilize the state of the brain at a given time:

— Computerized axial tomography (CAT) is an X-ray device based on the differential X-ray absorption of tissues, which increases its effectiveness over classical radiology. It detects anatomical aspects, yielding two-dimensional images of the shadows cast by radiopaque cerebral structures.
— Magnetic resonance (MR) and nuclear magnetic resonance (NMR) generate quantitative data on biochemical variations in the brain tissue. They allow for the visualization of the encephalon, along with other organs, by detecting the resonance (i.e., the response) of hydrogen atoms or other elements present in the tissue in question. After being aligned by a strong magnetic field, the hydrogen atoms are diverted by a transmission of electromagnetic waves; these atoms emit radio signals detectable through image processing. The sociologist of science Amit Prasad explains: "MR images are computer-generated visual reconfigurations of physical data such as the relaxation times of hydrogen atoms that are found abundantly in the body."[41] He describes how, "unlike other medical imaging technologies, which depend on a single parameter to produce images of internal parts of the body,

> MRI [magnetic resonance imaging] can use multiple parameters such as relaxation times, proton density, or diffusion of blood or other fluids for image production."[42] For the person in the scanner, the processing experience is primarily auditory. The rhythmic sound of MRI is generated by a gradient pulse. The radio-frequency emissions of the transmitter coil, which surrounds the body part under examination, interact with the subject's tissues to produce the MRI signal. In an MRI scan, signal is expressed as voxel or pixel brightness and noise as the random difference in pixel values produced by the subject. It is necessary to filter out the noise, as it interferes with the reading of the image. The signal-to-noise ratio (SNR) of an MRI image can be calculated by dividing the average signal intensity of the tissue by the standard deviation of the background noise. The topographical resolution of the image is high, and the scans are performed quite rapidly; the radiologist can alter the shades of gray dynamically to locate the pathology in a process known as windowing. Comparisons can also be made by changing the gray contrast in specific areas of the image; this process is known as leveling.[43]

Functional techniques visibilize the cerebral processes corresponding to mental activities.

- Functional magnetic resonance (fMR) and functional magnetic resonance imaging (fMRI) rely on high-speed sequences that map resting- and active-state regional blood flow in the brain in response to sensory, motor, and cognitive stimuli. They make use of the magnetic properties of the oxygen-transporting hemoglobin in the blood; these properties vary based on the amount of oxygen transported. Such variations, which correspond to functional variations in the brain, are detectable, making it possible to identify the structures active in specific brain functions: more oxygen being burned indicates increased activity.
- Diffusion tensor imaging (DTI) maps the brain's white matter by measuring the diffusion of protons in water.
- Positron emission tomography (PET) detects positrons emitted by radioactive isotopes administered to the test subject through the injection of a radioactive solution. It is therefore an invasive procedure.

It registers the circulatory and biochemical events in the encephalon underlying cognitive, emotional, and sensorimotor brain functions, making it possible to observe the functions themselves. This allows for a direct quantitative assessment of regional physiological processes in the brain. Researchers can monitor metabolic activity while the subject is performing the task. The image of the brain activity relative to the task under study is generated by a process of subtraction: the PET image of the patient's resting-state brain is subtracted from the image of the patient's brain during the execution of the task. It takes several minutes to obtain a single scan. PET is among the most complex, expensive, and deeply interdisciplinary sets of technologies available to brain imagers. Anne Beaulieu lays out the stakes: "In the 1980s, a debate around the interpretation of PET scans was fought out between the quantitative and qualitative (visual) approach to PET. . . . One side claimed that clinical PET can produce representations that can be understood visually; researchers insisted on a quantitative measurement and a quantitative evaluation. In this debate, PET scans were seen as highly processed data—involving normalizations, statistical tests, and the use of complex models typical of physics research."[44] The image is produced by a computerized translation of the algorithms that record metabolic neuronal activity. The familiar yellow-red color of the image results from a postproduction stage and indicates an area of increased brain metabolism.

— Single-photon emission computed tomography (SPECT) requires that the test subject be administered a contrast agent and is therefore invasive. It records brain activity by measuring metabolism.

— Electroencephalography (EEG) is a technique for studying the functionality of the brain that records bioelectric activity by means of electrodes placed over the scalp, ranging in number from 19 to 64, 128, or even 256. Electrodes record the electrical activity of groups of neurons with specific traits (pyramidal with apical dendrites perpendicular to the scalp). The activity is translated into brain waves on the screen thanks to an amplifier, which converts the signals from analog to digital. By means of source analysis, thus applying algorithms that model the brain, conceived of as a sphere, it is then possible to generate pixelated two- or three-dimensional images that show the underlying cerebral activations.[45] This technique is playing an important role in the growth of neurogaming, as the media scholar Ksenia Fedorova notes, because "scientific research shows that EEG can be used to decode motor intentions from a 'sender' brain that can be delivered as commands via transcranial magnetic stimulation (TMS) to the motor cortex of a 'receiver' brain, making the receiver involuntarily perform a movement imagined by the sender."[46] People play with interfaces and the result is a more ludic version of the out-of-body implants mentioned previously.

— Magnetoencephalography (MEG) records the magnetic fields produced by the electrical activity associated with brain activity. It records fluctuations in neuronal activity occurring at too high a speed for the resolving power of fMRI and PET but lacks the spatial resolution of fMRI. Like EEG, it measures neuronal activity directly, whereas fMRI and PET detect changes in blood flow and metabolism.

The iconic value accorded to brain images is a symptom of our addiction to pictures, which brain imaging raises to a higher level. We could account for our tendency to look at images by arguing that we, as humans, are primarily visual rather than tactile, olfactory, or auditory beings. But to do so would be misleading, not only because graphs, digits, signs, and diagrams are no less visual than images, but also because "seeing is not the ordinary technique of investigation" in this research field.[47] Brain imagers are apparatuses that not only overcome the skin/skull boundary but interpolate our sensorium, whereby they serve, master, and fabricate our visual environment. They produce a special kind of visuality that is in no way more accessible, more transparent, or less reductive than mimetic portraits, on the one hand, or abstract diagrams, on the other. Of course, some of the imaging techniques considered here are also used in nonneurobiological research to compensate for limitations posed by human eyes and their embeddedness in the head (CAT, for example, is deployed in archaeology to extract information about the contents of closed sarcophagi and to reproduce it by means of a 3-D printer). Digital photography, MRI, and CAT were likewise instrumental for the Visible Human Male and Female Project to develop digitized whole-body depictions of a man and a woman, seen from under the skin. The goal of these renderings is to provide a gold standard for comparing human anatomy and to obtain an otherwise unavailable integral display.[48] However, three features are specific to in vivo brain imaging. First, it is four-dimensional: it entails the three spatial dimensions of the organ plus a factor of time, and the activity shown is not contemporaneous with either the comprehensive scanning or the viewing, although the picture creates an impression of instantaneity with the former and synchronicity with the latter. To question this would be to destabilize the iconicity of brain images. Second, its reading is predicated on what is known as a reverse inference, a backward interpolation "from the presence of brain activation to the engagement of a particular cognitive function."[49] But reverse inference promotes a logical fallacy and is therefore not deductively valid. As neurologist Giandomenico Iannetti, a critic of the process, clearly explained to me: "Even if we identify a fairly selective signal for a condition A versus a condition B, there is no guarantee that signal (i.e., that neural activity) is useful or will contribute to our understanding of how the activity of the nervous system 'produces' mental state A."[50] For example, the claim that an emotional or cognitive experience—such as social pain—"triggers the same brain activity as physical pain is justified if, and only if, this brain activity is exclusive to the perception of physical pain."[51] The problem lies in the inadequate consideration given to the hypothetical clause. For the reasoning to be valid, the argument would need to be formulated abductively, proceeding from effect to cause and explicitly positing a link as one possibility among others. Yet conjecture is raised to conclusion, much as in neuroaesthetic studies, as I mentioned in chapter 2, hypothesis quickly becomes thesis. This strips away the falsifiability criterion adopted by science to ensure its own reliability: a theory must be falsifiable to be scientific. It is thus highly unfortunate, albeit unsurprising, that published experimental research does not include failed trials and unclear results. To acknowledge them would be to destabilize the objectivity of brain images. Third, as Fedorova points out, "Visualization of mind activity is grounded in how that activity itself is framed as an image. Yet, the only access to that activity is interpretation of the electric [or other kind of] signals that in their nature are dynamic. The image of a mental activity at stake is fluid and is based on continuous comparison of discrete states."[52] To recall this would be to destabilize the pictorial status of brain images. Owing to their opacity, this ensemble of images, when displayed in scientific publications and popular media, creates a double bind for both scholars and lay onlookers, regardless of critical caveats from the likes of Gallese, who maintains that "if brain imaging is not complemented with a detailed phenomenological analysis of the perceptual, motor and cognitive processes it aims at investigating, and, most importantly, if brain imaging results are not interpreted on the basis of our knowledge of clinical neuropsychological cases, this approach loses much of its potential heuristic value."[53] The two clashes mentioned earlier return to the forefront here. Brain images are viewed as if they originated from our real, natural, organic brain—which they, in turn, present to us—yet their digital status renders them totally alien to us, for they produce pictures that do not represent the world they claim to resemble. They are really afterimages (akin to those generated by the Spellbinder app, which uses algorithms to match digital camera-phone pictures with others stored on a central server, including shots taken under different lighting conditions). Rather than mediating between us and our appearance, they surreptitiously draw us into a new, uncanny scopic regime. As Joseph Dumit aptly states, "the image's apparent picture-like status and manufactured objectivity threatens to overwhelm its interpreted nature."[54] Such a threat is rooted in an antagonism between the analog-psychological personal mind and the digital-neuronal human brain, which, as I will discuss presently, fuels an "iconoclash" in our understanding of the brain.

Despite advances in technology and the very idea of a human fMRIzation, we are living simultaneously under contradictory epistemes. A representational attitude conflicts with nonsemantic procedures. It is not just a matter of understanding that "the space enclosed by the skull has been penetrated" and its contents, once extracted, are like a fish out of water.[55] What I mean to underscore is the epistemic short circuit between an increased exposure of the subcutaneous body and imaging apparatuses whose renderings are as yet unable to instruct us in how to look at them, the way perspective painting did by means of the focal point. Are they really to be trusted, or would we do better to question them? Are we to blame for overreliance on or excessive skepticism toward them? Is it necessary or advisable to decide one way or the other? We lack the skills for discernment and therefore experience what to me bears the hallmarks of the crisis Latour calls iconoclash, a concept he elaborated in 2001 to describe the ambivalence we feel upon witnessing others touching human-made images. Why are they handling them? As they are visual matter, are we not better off limiting ourselves to speculation from afar?

Iconoclash is "what happens when there is *uncertainty* about the exact role of the hand at work."[56] What I propose is that an analogous dilemma comes into play around the performance of the observer's eye, whose biological materiality has been perverted, making it a hybrid of a telescopic hand and a floating look. Brain-imaging techniques destabilize the very agency of human vision insofar as no immediate "seeing" is performed. De Vos explains: "In brain imaging technology the observer is technically incorporated into the machine. In other words, the measurements are made, the data is gathered by teleporting, as it were, the gaze of the observer into the scanner."[57] Our personal gaze is not simply embodied by the camera, as is the case in photographic and filmic practice.[58] It is manufactured by the technologies.

The medial nature of the gaze should not be misconstrued as neutralizing its nurtured agency, not least because statistical knowledge plays a part in the researcher's mode of looking, which scans for activated areas, and epidemiological knowledge intervenes in the medical examination, which looks for pathologies. This gaze is highly situated. What we gain is a surplus of visual presence, albeit of a different, fabricated nature. Prasad conceives of this visual regime as a "cyborg visuality,"[59] entailing a continuous development of new imaging techniques, whose aftereffects comprise a reconfiguring of the brain's visual appearance to exclude signals such as those from certain kinds of tissue (e.g., fat).[60] Dumit believes that the PET imager fabricates "a particular kind of humanoid: a cyborg whose experience of vision includes the physiology of the brain as witnessed through PET scanning."[61] In my understanding, the image at the center of this new cyborg scopic regime—the outcome of such a brain-machine interaction—presupposes four interwoven factors of disorientation: the imaged brain, the imaging of both operator and onlooker, and the computer systems involved. This regime is collective, heterogenous, and distributed over time and space. To be clear, this cluster is not exclusive to the production and spectatorship of brain images. Image making has, since time immemorial, brought together the power and the vulnerability of the maker, the materials and tools, the model, and the recipient. What is specific here is the nature of the making. To posit a cyborg visuality is to pervert the conception of visuality as the agency that characterizes our most prominent sense from birth: sight, which is unidirectional and not retroactively affected by what it sees, and which establishes the world in front of us and at a distance (unlike hearing, which is effective at 360 degrees and establishes proximity). The "look" of brain imagers is acoustic, magnetic, rather than visual. This change is instantiated in Marc Didou's bronze sculpture *Eco* (2004), consisting of two specular heads: "Eco translates the data produced by MRI into a corporeal, tangible reality. Yet this 'objective' reality reveals itself to be depending on the perceptual movements of the viewer's body."[62] From the standpoint of our inquiry, we can either accept Merleau-Ponty's claim from 1960 that "thinking deliberately reduces itself to a set of data-collecting techniques which it has invented"[63] and acknowledge that we have to relinquish something of our inherent visual agency, or we can argue that image making refashions self-awareness by negotiating between our natural-born capacities and the environmental apparatuses in which we enact them.

## THE BRAIN-IMAGE MILIEU

We have just seen how the mimetic conception of images that represent by way of resemblance, by referring to whatever serves as a model for figuration, and that produce copies comparable to their originals, is ultimately unsettled by digital media in general and by brain imagers in particular. The visual approach long dominant in the West has emphasized the realistic rendering of surface features of both things and living bodies to such an extent that, for Roger Fry, the progressive discovery of the modes of figuration based on likeness directly sums up the entire history of Western art.[64] Present-day image making disrupts what I would term the *quotation from the visible*. Of course, we could cite precedents for this prospect of disruption and call Fry's understanding into question. But I will resist rehashing Oscar Wilde's claim that there was no fog in London before Whistler painted it and Picasso's conviction that Gertrude Stein would someday resemble the portraits he made of her. In short, I will avoid discussing the artifactuality of image making as such, regardless of the Western drive toward mimetism, for this very mimetism is presupposed in the brain's iconoclash. Goodman formulates a strategy for understanding the link between images and "what they are images of" that dispenses with ontological presumptions and could be instructive for addressing the brain's highly mediated depictions. He observes that a painting must be a symbol of the object it is supposed to represent, and that this symbolic relationship results not from resemblance but from denotation, which in turn is independent of any similarities in visual appearance.[65] "A picture must denote a man to represent him" as a concrete individual, Goodman maintains, but "to be a man-representation," understood as an imaginary man, it "need not denote anything" real.[66] The tapestry of the *Unicorn in Captivity* from the series housed at the Metropolitan Museum of Art Cloisters represents a unicorn without denoting it, since unicorns do not exist. Conversely, that the face of the man with the blue sleeve in Titian's *Portrait of a Man* (ca. 1510) belongs to Ariosto, and that the painting therefore denotes the poet, does not alter the fact that the representation is still a figure-of-a-man—a specific type of figure. Hence, if a figure represents "a man as a man," then Ariosto qua poet is a figure-of-a-man as such, which also denotes and specifies the man as Ariosto. This kind of purely symbolic approach, however, would ultimately lead to a dead end, for it presupposes a realm of visual images already at hand, to be matched with possible referents.

The arrival of the neuron in our visual culture suggests that the rationale of art history as a "progress toward visual truth,"[67] or toward the faithful visual reproduction of the presently visible world, is being radically subverted. It is not just about the kind of dislocation of visual agency that computational photographer Steve Mann enacted in 1998, long before the advent of smartphones with video-calling capabilities, when he walked around linked to a web-accessible video camera so his wife back home could help him pick fruit at the supermarket. The delocalization and the spatial distribution of the gaze, as well as the dislocation of cerebral capacities away from the brain's container, are by now highly integrated into our lives. It is therefore not about the publicness of personal insight but about the paradoxical metamorphosis of the unvisual into the iconic. To be sure, if the twentieth century was the century of art for art's sake, our current time is that of the picture for the picture's sake. Stafford's plea that "images must finally make their exit from Plato's Cave"[68] has been answered. Despite the fact that "what is put under the scanner is . . . a psychological *theory*";[69] that, according to neuroscientist Max Bennett and philosopher of mind Peter Hacker, "PET and fMRI can scan brains, but not concepts and their articulations"; and that "neuroscience can investigate synaptic connections, but not conceptual ones"[70]—in sum, despite the incommensurability between neuroscientific knowledge and the visual by-product that conveys it—the following is indisputable: brain images attract a manifold spectatorship, and the trained gaze is no more cunning than the naive one. Both are nurtured by the same image culture in which pictures enact a reality effect and do so to such an extent that we are all lay onlookers playing neuroscientist to ourselves as if we knew the brain we were looking at.

The epistemic dilemma in which we find ourselves has been thematized by the whole spectrum of the visual arts, from drawings to installations, encompassing a variety of materials, since before the advent of the neuroaesthetic narrative. On the one hand, in 1973, the Italian artist Ketty La Rocca, trained as a radiologist, created *Craniologia*, which features an open hand superimposed on an X-ray of her skull with the handwritten word "you" appearing over the face to render its features. On the other, in 1998, Justine Cooper underwent six hours of MRI scanning to produce hundreds of images for use in the audiovisual animation *RAPT I*, the first MRI artwork to attract widespread public attention. While La Rocca displays a mundane portrait that merges traditional tokens such as the death mask and the *memento mori* and directly interpellates its beholder, Cooper penetrates her own living body, taken as an acoustic site that grants the beholder entry to a transcendental dimension. *RAPT I* lends character to the anonymity of the Visible Human Female Project, which it resembles from the standpoint of the devices used. Its rationale dismantles what W. J. T. Mitchell criticizes as "the fallacy of misplaced concreteness," whereby technologies are construed as nothing but tools for defining our material and intellectual performances.[71] It confronts us not with the deterministic power of the apparatus but with its performative capacity. In the crowded contemporary art scene, such a capacity is often associated with BioArt, which offers the latest proof of the deep-rooted bond between artistic production and scientific research and, when broadly defined, is taken to include images of the brain. In *Portrait of Sir John Sulston* (2001), Marc Quinn depicts the subject, a Nobel laureate for research in genetics, using the latter's genomic sequence extracted from a semen sample. To me, this work belongs to Quinn's series of perceptual portraits, as do his likenesses of John Coakley Lettsom, founder, in 1773, of the Medical Society of London, whose face is shown in *Founder Member* through the written list of members present at the society's first meeting, and Matthew Baillie, whose face appears superimposed on the cover of his 1793 text, *Morbid Anatomy*.[72] In *Identity Analysis* (2003), a stainless-steel sphere, the German artist Helga Griffiths has etched her own genetic code, converted into a series of small rectangular holes, through which a powerful light originating at the sphere's center filters out, to be reflected on the bodies of the visitors passing through it and multiplied by the mirrors on the surrounding walls.[73] Such an extension of the imaginable has pushed the limits of visual production, blurring the boundaries between artistic genres and prompting reenactments. We might consider a pair of case studies: the stark contrast of lights and dark patches in *Under the Blood* (2002)—a thirteen-minute documentary by Jordan Baseman covering a seventy-five-hour open-heart surgery and transplantation procedure at Papworth Hospital Heart and Lung Transplant Unit in Cambridge—has Caravaggio-style works as referents;[74] similarly, the performances of the Italian artist Franko B[75]—who in *I'm Not Your Babe* (1995–96) cut his flesh, letting the blood run into a pool at his feet until he passed out—have embodied depictions of Saint Sebastian the martyr[76] or Bellini's *The Blood of the Redeemer* (ca. 1460), where Christ appears leaning on the Cross with a red spurt flowing unstinted from the wound in his chest to a cup held up by a kneeling angel. The current extension of the imaginable carries over from portraiture to landscape. Thorbjørn Lausten's *MAGNET—Visual Systems* (2008) is a total environment that displays both meteorological and geomagnetic data. The former come from the Meteosat satellite positioned over the equator, which displays them every fifteen minutes, and the latter are received continuously from four detection stations in Norway and the Barents Sea. The data are retrieved and updated via the internet. The work is intended to highlight how technology functions as an extension of the senses: "In many cases, we are dealing with such a large quantity of data that the phenomenon in question can only be understood as a visualization and that it cannot be misconstrued as being an image of the phenomenon. . . . We cannot say that what we call reality is something entirely external to ourselves; we must include our neurological system and our (artificial) constructs."[77] Expanding the very notion of the living organism is central to works such as those by BioArtist Eduardo Kac, a pioneer of transgenic art, who in 2000 created the now classic *Alba*, a bunny that appears green owing to injection with the fluorescent gene of the jellyfish *Aequorea victoria*. In 1999, Ars Electronica 99 commissioned Kac to create *Genesis*, which used biology in conjunction with the internet to synthesize a gene based on the Morse code translation of the Genesis verse: "have dominion over the fish of the sea and over the birds of the air and over every living thing that moves upon the earth."[78] The Morse code was then translated into DNA base pairs, following a conversion principle specially developed by the artist, and the gene injected into bacteria. From there, anyone could act on the culture—even remotely via the internet—by activating an ultraviolet light in the art gallery, to trigger a real biological mutation in the bacteria. After the public exhibition, the DNA was translated back into Morse code and then into English, thereby modifying the order of the letters in the Bible verse. "We live in an era of trans-everything," remarks Stafford in reference to Kac as well as to Stelarc.[79] Her claim could easily be extended to the ensemble of case studies just mentioned. A quick reference to Damien Hirst, anecdotal though it may be, serves to illustrate. In 2005, the artist summoned emergency responders to his house after his wife suffered a fatal heart attack. Soon afterward, he "became the victim of federal prosecutors who discovered that Hirst had created in his home a space that resembled a biological lab, in which he was growing mysterious bacteria, all part of the careful scientific research he was doing for his creative 'bioart' work."[80] The idea of an increasing indifferentiation of media and matter, bodies and technologies, could paradoxically integrate brainhood into the posthumanist attack on anthropocentrism, for the human neurobiological system is as material as any other organic body. But insofar as BioArt remixes organisms and draws new taxonomies of life-forms, to me it does nothing but complexify the notions of nature and naturalism, the two highly Western cultural conceptions whereby human beings alone are endowed with an inner life, despite being linked to the continuum of nonhumans with which they share material properties,[81] but also kept firmly within the representational paradigm. Neglecting the specificity of the apparatuses that visualize life processes would open us up to the kind of situation I warned against when discussing the methodological pitfalls of neuroaesthetics, stating that all cats are gray at night.

The same blurriness is at play in the constellation of brain-oriented images. Is the brain a focus or a site of cognitive processes? To shed light on this question, I propose to classify these images into two typologies. Those belonging to the first retain a semiotic link to the brain as an organ and fit with the broader artistic approach observable in literature, film, and theater that makes it and its functioning a topic of inquiry, a narrative or figurative resource.[82] Jan Fabre, who has devoted a large part of his production to the brain as a cultural artifact and tackled it across several media, from drawing to theater, exemplifies this trend at its best. Two of his more recent works deserve mention: *Pietas* (2011), a marble reenactment of Michelangelo's *Pietà* set between four enormous white marble cerebellums where the dead Christ holds a brain, and the fifteen-minute video *Do We Feel with Our Brain and Think with Our Heart?* (2014) from the exhibition of the same title, which shows Fabre interviewing neuroscientist Giacomo Rizzolatti about the role of mirror neurons in cognitive and emotional processes. The interlocutors were filmed wearing EEG headsets. But the brain's assimilation, as a design or shape, into (visual) art production is actually attested by a variety of pieces, all of which emphasize a direct shift from the face (or head) to the brain as an index of identity. *Self-Portrait* (1991) shows Helen Chadwick holding what is implied to be her brain in her hands, notwithstanding the impossibility of such a feat.[83] In the photographic collage *Memory Picture No. 12* (1992) from *The Anatomy Lesson*, Joyce Cutler Shaw obscures the oval of her own face with an MRI of the coronal section of someone's brain.[84] Karen Norberg's *Knitted Brain* (1994), using fabric, re-creates the cerebral anatomy down to its minutest detail.[85] Susan Aldworth conceived *Cogito Ergo Sum 3* (2002), a series of twenty digital prints based on fMRI scans of her brain taken at Queen Square Imaging Centre in London, for which she altered the scale and manipulated the scans themselves to include words tying the medical images to her everyday experience. Her own reflection is instructive: "What is a brain scan? Brain scans are precise scientific photographs but they are incomplete records of a person. I am fascinated by their ambiguity: the gap between what they do show (the physical structure and function of the brain) and what they don't—the self. Brain scans have a specific meaning within a medical and scientific context, but this changes when we consider them as images of someone. For me, they signpost the interior self."[86] Constance Jacobson blends watercolor and ink to make cerebral and neuronal shapes in the series *Brains* (2008), reenacting the work of Ramón y Cajal.[87] Marta de Menezes, for her *Functional Portraits*

(2002), used fMRI to record her own patterns of brain activation while she focused on drawing and those of scientist Patricia Figueiredo, who played the piano. Proclaims the Portuguese artist: "Here is an instance where it becomes possible to create art with mere thought."[88] It is a nonfacial portraiture that hybridizes to the highest possible degree what physiognomy set out to do. In *Warm Glow* or *Fabric MRI Bill's Brain* (2009), Marjorie Taylor decorates a wool rug with twenty-three horizontal sections from an fMRI of a human brain, reproduced roughly but unmistakably. The series by Dario Ghibaudo *I peccati capitali sono otto* (The deadly sins are eight) (2011), displayed in the fifteenth room of his *Museo di Storia Innaturale* (Museum of unnatural history), consists of eight hanging rugs fashioned after Arab prayer carpets; each has a brain portrayed at its center with certain parts colored, as if representing the specific sin whose name is woven into the fabric (albeit without reference to a possible correlation between moral behaviors and cerebral areas). In *Self Portraits 4* (2012), Angela Palmer maps the sections of the brain onto individual sheets of glass, which are layered together to form a sculpture. The result is a 3-D image, viewable from certain angles only.[89] Jane Prophet's *Neuro memento mori* (2015)—inspired by a nineteenth-century wax model of a female head representing life and death—is a 3-D-printed polymer sculpture of data from an MRI, taken while she meditated in the scanner. Michela Martello's *Grandma's Mirror* (2017) juxtaposes, on a linen sheet, a bright pink drawing of a brain with an antique lace featuring imagery from the ancient world. The idea of brain tissue as lacework was already on display in Michel Thiebaut de Schotten and Benedicte Batrancourt's *Lace Brain* (2012), whose abstract cerebral illustration, created using "diffusion images, photoshop color filters, and a final filter called percolator," serves as the official cover image for the Organization for Human Brain Mapping.[90] In *Resonance Punctuated CXLIV* (2018), Laura Jacobson uses ceramic and stains mounted with epoxy on painted wood to display cross sections of human MRI taken from the axial direction (i.e., horizontal to the upright body); as she explains, "the dented ceramic comes to resembles bio scapes marked by culture."[91] Mental images take the stage in the art of Pierre Huyghe, starting with *UUmwelt* (2012), which was produced at the Kyoto lab run by the same Kamitani mentioned at the beginning of this chapter. Subjects under an fMRI scanner were given verbal descriptions or shown pictures of living creatures, prehistoric tools, machines, codes, and artworks; the data were then processed via an AI neural network—software mimicking the modus operandi of the human brain—which "learn[ed] and progressively reconstruct[ed]" the images by matching them with pictures taken in the real world.[92] Huyghe's project continues with the aptly titled *Mental Images* (2020).

These examples of artworks of such diverse workmanship suggest that the brain has gained not only a figurative recognizability but also its own visual status. On the one hand, it is a visual artifact that we can look at and handle and, on the other, it looks at us, and its vision is what we put on display. Well suited to this imaged brain is the quality of hyperrealism, which, according to Stafford, comes about when artistic activity forces its material "to become more than it was when it existed in the real world." She defends these views in a dialogue with Bredekamp, who, for his part, defines hyperrealism as everything that "continuously ironises the monopoly of either seeing or grasping."[93] Indeed, the kinds of artifacts I have mentioned, which display images *of* the brain or *via* the brain, are conceived of much as they would have been before the disruptive effects exerted by audiovisual and spatiotemporal scanning on the inner/outer dualism. There is no subversion of the corresponding spectatorship. They are looked at; they can be held, hung up, or projected. They have a front and a back, top and bottom, left and right that correspond to the bilateral symmetry of the human body. Notwithstanding the diverse materials and technologies involved, what John Michel Krois calls our "pictorial competence" remains in force: "[this] does not involve postulating the existence of an innate grammar," or what to my mind is a special aesthetic knowledge capable of reconstructing the spatial information conveyed by the image; it means "only that an intelligent organism has a body with motor- and sensory perception" that resonates with the visual environment in which it is embedded.[94] All these works, artistic though they may be, still belong to a world in which we look at pictures that externalize our hidden self, the way European ladies saw their figures reflected in the *psyché*, the free-standing full-length mirror on a pivoting frame found in virtually every middle-class home in the nineteenth century.

But when self-awareness is at stake, when our own body's cyborg status is acknowledged, when we experience our body/brain system as already entangled with the "outer" world, a second typology of brain art comes to the fore. Here the data have the potential to take any form thanks to algorithmically established correlations,[95] and the brain the onlooker encounters out there is not the reified, objectified him or her but the result of an environmental nexus produced by image-making apparatuses that, as I have stated, intrude upon and expand the sensorium of both artist and spectator, producer and experiencer. What constitutes *that* brain image is already the performative space of the real lab. The interplay of the four factors of disorientation mentioned previously makes up the actual medium of brain images. The media theorist Aud Sissel Hoel casts further light on the matter: "In its mediating role, the MRI machine [and the like] does not provide a transparent window into a reality that precedes the imaging process, but nor does it fabricate distinctions. It operates, rather, by instituting a mixed, crossover environment, where technical and natural elements and forces come to intra-act and mutually influence each other."[96] If this is true of all postphotographic production—including, for example, the thermal imaging technology used by the French photographer Antoine D'Agata during the COVID-19 pandemic to document the liveliness of anonymous people—it is even more so of brain imaging. I would argue that any resistance to the realistic determinism conveyed by the icon of the cerebral Self derives from the fabricated image as such, which is grounded in a process of visibilization. I therefore endorse De Vos's claim that "it is only from there . . . that the . . . reductive potentials of the technological gaze (and their ideological) bearings can be discerned and eventually criticized."[97] The mediatedness of the image makes it possible to suspend the transitivity of the scholar, the onlooker, and the person whose brain is generated so that the voyeurism of brain imaging can be questioned. I thus contend that brain imagers do not necessarily produce iconic images. They also produce instances of the technically mediated cyborg visuality that enhances self-awareness. Pieces such as the following enact an artistic feedback loop of this sort.

Andrew Carnie's and neurobiologist Richard Wingate's installation *Magic Forest* (2002) invited visitors to walk through a forest of floating trees, at once oneiric and familiar, which were actually recordings of brain cells transmitting signals to dendrites.[98] Darij Kreuh and Janez Janša's performance *Brainscore* (2003) blended virtual reality technology and brain signal processing to create a kind of neurogame in which two operators, acting through avatars, could communicate by brain waves and eye movements. Mariko Mori's *Wave UFO* (2005) engaged three viewers at a time: "Each viewer gets electrodes attached to the head, just like an EEG machine. The brainwaves of the viewers are transmitted and projected on to a screen. This will show six orbs for two of each viewer's left and right brain hemispheres. A waving line shows the facial movements of the viewer."[99] Artists such as Marco Donnarumma and Dmitry Morozov, known specifically for their work with biofeedback, make art based on physiological data, from brain waves to circulatory rhythms to muscular contractions. In *The Escalation of Mind* (2012), Morozov turned the human brain into a control voltage generator, monitoring in real time the cerebral outputs, emotional state, and facial expressions of an actor who quoted passages from Herman Hesse's novel *The Glass Bead Game*. The signals served to generate an audiovisual environment. In Ursula Damm's interactive installation *Chromatographic Ballads* (2013), visitors outfitted with EEG devices could watch as their brain signals, equated in the exhibit with unconscious reactions, altered images of public spaces. In 2014, the prolific artist and neuroscience researcher Suzanne Dikker collaborated on *Measuring the Magic of Mutual Gaze* with Marina Abramovic, who adapted it from her performance piece *The Artist Is Present* (2010). The experiment-like setup called for two participants wearing EEG devices to engage in a mutual gaze for thirty minutes while their brain waves were recorded and visualized. Synchronization of the brain waves would trigger a lightning animation, thematizing human connectedness. Greg Dunn and Brian Edwards's work *Self Reflected* (2014–2016) uses data from actual MRI brain scans in combination with a special technique known as micro-etching to depict neurons. Relying on anatomic maps and data-processing algorithms to ensure a realistic outcome, the artists worked against a reflective gold background, creating collections of tiny angled ridges that gather and reflect light. Movement by either the observer or the light source causes the image to appear animated, brightening or darkening in different areas successively to simulate the electrical activity of a brain looking at art.[100] Laura Jade's *Cerebral Nebula* (2015) is a life-size sculpture of the human brain, created from an MRI scan of a thirty-five-year-old male that was divided into cross sections, laser cut in Perspex, and hand-etched slice by slice with neural networks as a prototype for the artist's *Brainlight* project. Indeed, Jade is currently using an EEG headset to explore specific brain frequency bands—meditation (Alpha), focus and attention (Theta), and excitement/agitation (Beta), in addition to emotional states and facial expressions—all of which will be rendered as a light display within the brain sculpture. Her aim is "to experience a level of conscious mind-control over [her] brain's bustling electrical activity while simultaneously witnessing an internal feedback loop created by our constantly changing perception of the artwork."[101] This progression from a prototype to a fully realized project suggests that the two types of brain images I have proposed are not necessarily mutually exclusive but can integrate with each other. Ellen Pearlman's *Noor: A Brain Opera* (2016) channeled data from the real-time brain-wave monitoring of an actor into an immersive, interactive audiovisual performance based on the story of Noor Inayat Khan, a covert wireless operator who infiltrated Vichy France. Against a spoken-word account of the protagonist's life, the performer moved about the room, mixing with the audience, while her changing emotional states were transmitted and visualized over a series of displays.[102]

In the end, art and science do not exchange roles. As the art historian Martin Kemp, highly attentive to the affinities between the two domains, points out, an artwork always remains open to the viewer's interpretation without being able to compel its own beholding. The scientist, by contrast, while perhaps wishing to seduce with the prettiest visuals of brain activity, ultimately seeks to communicate an interpretation that can withstand counterevidence.[103] The image also engages us in different ways, as Gombrich conveys in an incisive reflection: "X-ray specialists basing a diagnosis on the faintest of shadows visible in a tissue, learn in a hard school how often 'believing is seeing' and how important it therefore is to keep their hypothesis flexible. The art lover adopts the opposite mental set. Unless he is a restorer, he may go through life without ever realizing to what an extent the pictures he loves are crisscrossed by subjective contours of his own making. If he were ever to strip them of these projections, merely a meaningless armature might well be all that would remain."[104] A change does occur, however, in that the mediation enacted by brain imagers requires a new understanding of the brain's iconicity, on the one hand, and of its performativity, on the other.

# 6

# BRAINS ON STAGE

In the first century BCE, philosopher Apollonius of Tyana asked his beloved pupil Damis of Nineveh to describe the nature of painting. Painting, said Damis, was a mixture of colors made for the sake of imitating visible bodies, for people are given to re-create what they see. He conceded, however, that to our eyes, clouds, too, resemble meaningful forms, because we discern in them the appearance of natural bodies. Apollonius ruled that imitation was twofold: "the one kind as an employment of the hands and mind in producing imitations, [this] is painting, whereas the other kind consists in making likenesses with the mind alone." But Damis rejected this proposal, arguing that if painting were "a faculty of making likenesses with the help both of mind and hand . . . we must regard the other kind as a department of that, since its possessor perceives and imitates with the mind, without having the delineative faculty."[1] That is, we paint by seeing. Imitation, then, can be understood as occurring in two moments. The first pertains to the perception of forms (whether two- or three-dimensional) and of actions, by processes that frame portions of the visible realm as salient. This moment is imaginative and defines spectatorship. The second pertains to the re-creation of those forms and actions. This corresponds to art production and is largely, though not necessarily, material. The two are not independent of each other. The ability to recognize visual patterns somehow participates in the image making itself, which does not need to be externalized to exist at some level. Onlookers internally simulate the creative gesture without being trained to perform it in the outer world. Against the background of our current neuroscientific knowledge of the embodied simulation process, which is predicated on the activation of the cerebral visuomotor mechanism whereby viewers imitate images and actions observed in real life or in depictions by executing them mentally, I regard the twofoldness of imitation as an image/imaged loop: there is a spectator in every *faber*, be it a potential maker, an actual craftsman, or a skilled artist. For the purposes of neuroaesthetics, I posit visual art—with the exception of virtual art, which disrupts the very notions of picture making—as largely the concrete outcome of a pragmatic insight; it results from the bodily deployment of a personal mimetic faculty that is mental in nature and only afterward enacted on media. Indeed, what qualifies a visual configuration as meaningful is not primarily its provenance (i.e., as handmade or not) but its recognizability—as determined by our mental or material processing—which, in turn, attests to its origin.

Such a claim raises more questions than it answers. Several research approaches, different among themselves, address it from a performative standpoint; that is, they dispense with the representational fallacy, which takes artworks to be epiphenomenal outputs of the human mind or impressions of standard neurobiological mechanisms on matter. One is that of Tim Ingold, who contends that humans learn by making and know what they know because they have made it, so that any distinction between the skill required for things, crafts, or artworks is blurred.[2] Another is that promoted by Vittorio Gallese, the main theorist of the embodied simulation process. He maintains that the fictionality distinguishing things from artworks corresponds not to their social status, production techniques, or stylistic features but to the ability of beholders to re-create them; as he points out, the "'as-if' mode of embodied simulation appears to typify not just our appreciation of fictional worlds, but all forms of intentional relations, including those characterizing our prosaic daily reality."[3] Fiction, then, is to be understood in its original Greek sense, as modeling, sculpting, giving shape. A third line is exemplified by Maria Alessandra Umiltà and colleagues' discussion of their experimental studies on Franz Kline's abstract paintings, which takes the mental nature of the mimetic faculty to actually reinforce the fictional gap. These researchers, starting likewise from the visuomotor mechanism, compared the cortical areas activated by the observation of the artist's brushstrokes, which are derived from his motor acts, with those activated by the observation of graphic patterns that are similar yet free of dynamic components, such as drips of paint, blurred contours, differences in pressure on the brush. They concluded that looking at handmade paintings produces a clearer and significantly greater cortical activation than looking at photoshopped stimuli.[4] These approaches all foster the idea that reality and art, facticity and artifactuality, participate in the same mental and material world, and that aesthetic specificities in the responses elicited by the two domains are the function of what I term a *performative gradient matter*, which draws out a graduated visual agency. No radical heterogeneity intervenes among the human, the environment, and the artifact. Rather, a family resemblance is in play. From the observer's point of view, the fabrication of a window, a window, a painting of a window, and a rectangle all align. Visual marks are inscribed in abstract art just as abstract art is implied in mimetic art, since all three entail an original gesture: the performance whereby the pattern could be made.

German aesthetics has dealt extensively with the corporality of visual processes. Robert Vischer discusses the tactility of sight in "On the Optical Sense of Form: A Contribution to Aesthetics" (1873). Citing Gustav Adolf Lindner, he suggests that "seeing is a 'more subtle touching at a distance' ";[5] it is an action in which the gesture is not suppressed but remains latent, even though visual acuity—capable of observing faraway bodies—takes precedence over even the subtlest finger, and the hand, conversely, is what gets close to images. The hand is bound to the eye, as its jointly responsible accomplice, and vision is the hand's prosthetic extension. In "The Problem of Form in the Fine Arts" (1893), Adolf Hildebrand writes that the child develops its own visual representation of space by moving its legs and arms and using its eyes, which, though encased in the head, are likewise "moving."[6] Giacomo Rizzolatti and Corrado Sinigaglia bring the alliance between hand and eye to the foreground once more in *Mirrors in the Brain: How Our Minds Share Actions and Emotions* (2006), arguing that "the seeing which guides the hand is also (and above all) seeing with the hand, by which the object is immediately coded as a given set of *invitations to act*."[7] But whether this invitation actually triggers the *production* of visual forms, and whether there is also a virtual *faber* in every viewer, is a slightly different matter. It is one tackled in the 1926 text

*Affective Thought* by John Dewey, who, along with James J. Gibson, contributes to the theoretical backdrop of what I term *performance-based neuroaesthetics*, an approach in the visual arts whereby images acquire a performative capacity owing to the human mechanisms of embodiment, and are entangled with media. The two hold a position akin to that of Gombrich and Arnheim vis-à-vis visual neuroaesthetics. John Dewey's short essay—really a review—is among the texts that establish the American philosopher as a main reference point for the current biocultural approach. He argues that "unconscious activities are realities," that the spectator "clicks" before the artwork owing to engrained organic memories, and that "eye activities arouse allied muscular activities which in turn not merely harmonize with and support eye activities, but which in turn evoke further experiences of light and color, and so on . . . so that *integration in the object permits and secures a corresponding integration in organic activities*."[8] His arguments were developed in response to art collector Albert Barnes's *Art in Painting*, a treatise dedicated to Dewey himself (the two, being friends, greatly influenced each other).[9] Writing about appreciation in the (visual) arts, Barnes contends that "the artist, whether in paint, words, or musical tones, has embodied an experience in his work, and to appreciate his painting or poem or symphony, we must reconstruct his experience, so far as we are able, in ourselves."[10] An artist such as Henry Moore would agree: "The sculptor. . . must strive continually to think of, and use, form in its full spatial completeness. He gets the solid shape, as it were, inside his head—he thinks of it, whatever its size, as if he were holding it completely enclosed in the hollow of his hand."[11] Such a merging of materials, design, and bodily engagement was addressed at the beginning of the twentieth century by Jean d'Udine, alias Albert Cozanet, a French composer and art scholar worthy of greater attention, who based his proposal entirely on the synesthetic experience. He hypothesized that touch, having initially served to train the other senses, continues to mediate between them, and this mediation lies at the root of synesthesia. Adapting his reasoning to art, he proposed that "to each emotion, of any order whatsoever, corresponds a behavior, a bodily movement, and one only, and it is through this movement that the extremely complex synesthetic translation accompanying all artistic creation operates."[12] The gesture comes together in an aggregation of matter: the work. Though inert, the work can always reproduce the gesture, transferring it to other individuals and enabling them to feel the (e)motion in question. The artifact becomes like a fossil containing the "plastic form of a state of mind"; the creative move is reified but always ready to be reactivated.[13]

It would be worthwhile to undertake a genealogy of practitioners, in the visual and performance arts as well as in architecture, who espouse views in line with these, given the many interconnections that exist between their know-how and the realms of science and philosophy. As it happens, artists who are rehearsing strategies in order to find efficacious results develop a better-honed expertise than academics concerned with conforming their explanations to the most up-to-date findings. Regarding embodied simulation theory, it is indeed with some condescension that theater director Peter Brook famously credited "neuroscience" with "finally start[ing] to understand what has long been common knowledge in the theatre."[14] Such a long-standing awareness dates back at least two and half millennia to the time of Plato's *Ion*, where the rhapsode describes his hypnotic effect on the audience in term of two poles brought in contact by magnetic inspiration. In Western culture, over the centuries, various clusters of scholars and artists have expounded the power of bodily imagination. One exemplary line starts in the late nineteenth century with the Viennese physician Salomon Stricker's focus on subliminal muscular sensations, coupled with his ideas that spectators experience, in their muscles, the actions of others, and that all thinking is based on the inner perception of muscular dynamics. The French psychologist Théodule Ribot carefully studied Stricker's work and in turn became a vital resource for Konstantin Stanislavski, the drama theorist and director who developed the operational definition of emotional memory and taught actors to stimulate their own feelings through physical actions. Gallese himself acknowledges the interest aroused by the theatrical display when he writes that "through a dissection of the actor's behavior, [theater anthropologists] retrace . . . the totality of his expression to a multiplicity of levels of organization of the corporeal practices of movement," adding that "this constitutes a natural bridge of dialog with cognitive neuroscience."[15] Such a theoretical and pedagogical conundrum in which the habits of performer and spectator overlap is currently discussed in dance studies as "kinesthetic empathy"[16]—that is, empathy for the perceived motion. As a historian of the visual arts, David Freedberg clarifies that empathy, by contrast with sympathy, "should be considered not so much as an all-purpose account of a sense of understanding the emotions of others, but as an account of bodily engagement with their seen movements."[17] In his 1996 text, *The Eyes of the Skin*, the Finnish architect and architecture theorist Juhani Pallasmaa opposed the supremacy of opticality in his field and advocated "re-sensualis[ing] architecture through a strengthened sense of materiality and hapticity, texture and weight, density of space and materialised light"[18] by rediscovering the neglected senses. Some two decades later, he found in embodied simulation theory a counterbalance and effective antidote to "shallow rationality and reliance on the empirical, measurable, and demonstrable,"[19] arguing that neuroscience can offer a critical barrier against current tendencies to intellectualize architecture and our relationship with the environment. His perspective on the neurohumanities, then, is in no way connected with brainhood but rather echoes Gallese's viewpoint that "the functionality of the motor system literally carves out a pragmatic *Umwelt* dynamically surrounding our body."[20] As Victoria Pitts-Taylor similarly argues, "a mind that is embodied, potentially, can mean it is immanent (tied to the capacities and worlds in which it is enacted); relational (affected by its position to and interaction with other minds, bodies, and objects); affective (shaped by feeling and emotion); and situated (tied to specific places, needs, and circumstances)."[21] If, in the age of neuroculture, brains are nothing but metaphors for the Self, I view them as wielding a performative agency over personal identity. This agency, in turn, is grounded in the very cerebral plasticity posited as the organic medium for our entanglement with the environment.

The move away from ocularcentrism, or "visual brain imperialism" as Gallese calls it, which I discussed in terms of neuro-opticality in chapter 2, toward a multimodal conception of vision based on the dynamics of the motor system in an environment inhabited by artifacts and people inaugurates this new epistemic framework. The power of sight to capture objects in the distant, extrapersonal space was rooted in the egocentricity of perceptual activity, while the notion of enactive cognition, theoretically predicated on the interaction between humans and the surrounding world through which they extend, integrates the telescopic eye. In the shift from skin/skull envelope to body/brain dynamic, we have gone from thinking of vision as a slingshot to thinking of it as an integrated, dynamic field. This integration enables us to discard the solipsistic stance, which posits an autonomous subject, and in its place adopt the idea of performing, situated, prosthetic identities—perceptual, cognitive, emotional agents interacting in crowded, shared, hybrid milieus. Such a desubjectivation might seem paradoxical, given that the design of the neuroscientific experiments and methodologies requires researchers and participants to express themselves in the first person. Yet this first-person perspective speaks not of the Subject but of those empirical individuals whom we are when we say "I" in everyday life and who are potentially standardized ahead of time by a quantitative approach. What we might have to contend with, then, is an all-inclusive theory of the performative agency of our all-inclusive, expanding sight: a theory of fluid interpenetration, of *con*fusion among humans, artifacts, and apparatuses, that is divorced from personal and medium specificity and that therefore also disrupts the evolutionistic background generally implied by all neuroscientific knowledge. In this final chapter, I will thematize the implications of this epistemic perspective in relation to a decisive trend in contemporary artistic production.

## IMAGES AS PERFORMANCES

Knowledge about mirror neurons dates to the early 1990s.[22] According to the bottom-up approach that led to their discovery and the embodied simulation theory that Gallese later developed around them, the fact that the same brain area is activated when we perform actions or experience emotions or sensations as when we observe others performing or experiencing them constitutes evidence of a direct, prereflective, nonvoluntary understanding of others' actions, emotions, and sensations. It presupposes a nonconscious, simulative body/brain system that works by modeling things, agents, and movements. Simulation is activated during mediated and unmediated interactions with others and is influenced, in turn, by contextual, cognitive, and personal-identity considerations. "Through its plasticity and modulation," argues Gallese, "embodied simulation . . . might be also the vehicle of the projective qualities of our aesthetic experience, where our personality, cultural identity, the context, our mood and disposition, literally shape the way we relate to a given perceptual object, by means of the modulatory effect exerted on our simulation processes."[23] Such a straightforward and by now established definition, which rules out any reductionist approach, crowns the efforts of the Italian, Parma-based neuroscientific research group initially headed by Rizzolatti and made up of Gallese, Giuseppe di Pellegrino, Leonardo Fogassi, and Luciano Fadiga, to be joined later by many others whom we will encounter in the coming pages as the developers of decisive case studies. Their contributions to neuroaesthetics over the years belie the criticism that "while neuroaesthetics has accumulated important empirical findings, it lacks the analytic tools to interpret them in a way that is both theoretically sound and true to aesthetic experience from a first-person perspective."[24] Rather, if neuroaesthetics continues down this performative road, it could help dismantle the rearguard position that holds culture and biology to be separate, when really they result from human cognitive technology and in fact coevolve as media-mediated.

The seminal experiments on mirror neurons were conducted on a macaque. Researchers implanted several microelectrodes, accessible via a prosthesis, into its cortex (which is devoid of nociceptors, rendering the invasive technique painless), to stimulate as well as record cortical activity. According to the setup, the electric charge emitted by the "firing," or activation, of several single neurons would generate a tracing; the microelectrodes, in detecting variations of electrical potential, would also produce an intermittent noise with greater or lesser frequency. The researchers were investigating the F5 area of the macaque's premotor cortex, the region of the neocortex involved in selecting, planning, and executing actions such as grasping objects and putting them in the mouth. The neurons in this area respond to actions of the mouth and the hand, and the test was designed to study one neuron's activity over a full grasping action. As it happened, however, neurons fired loudly while the monkey was at rest and Gallese and Fogassi were taking a break.[25] What prompted the activation? Although at the time the scientists were caught off guard and had no ready answers, the knowledge they developed through subsequent inquiries, both empirical and theoretical, has been hailed by Ramachandran as equally relevant "for psychology [as] DNA [was] for biology."[26] Indeed, the researchers involved, while generally agreeing that neither the neuron nor the cerebral mechanism bears an actual epistemic agency, posit the mirror system as the neural basis for cognition, intention, action, representation, and ultimately, social interaction.[27]

The researchers succeeded in distinguishing between two types of neurons: "canonical neurons" and "mirror neurons." Canonical neurons fire both during the observation of an object that could be implicated in an action and during the execution of that action at first hand. They fire, for instance, when a monkey notices an apple that could be grabbed—that is, when it sees the round shape of an object that happens to be an apple—and when it actually grabs an apple. Different neurons respond to distinct types of things, which are classified based on considerations of manipulability (whether large or small, etc.). Bernard Berenson had already identified, in painting, "tactile values," which he ascribed to solid represented objects that were not merely imitated in a realistic manner but "communicated . . . in a way that stirs the imagination to feel their bulk, heft their weight, realize their potential resistance, span their distance from us, and encourage us, always imaginatively, to come into close touch with, to grasp, to embrace, or to walk around them," so that their pictorial "form" primarily indicates the tangible quality of the figuration.[28] A slight difference, decisive for our discussions ahead, arises here between what an item (e.g., an apple) is by itself, for others, and so on, and how it appears *to me*, how I picture it for myself. Any object, and any action, is salient insofar as it interpellates me. Indeed, mirror neurons fire when the subject executes an action at first hand and when he or she *sees* the same action executed by another; it makes no difference if the observed performer of the act is another monkey or the human experimenter, provided that the observer is physically equipped to execute the observed action. However, for the monkey the gesture cannot be an intransitive one, such as randomly shaking an arm, but must serve a goal involving the manipulation of an object. Each transitive gesture, then, is classified in terms of "grasping" mirror neurons, "holding" mirror neurons, and so on. Intransitive mirror neurons, which were discovered afterward, discharge when the monkey observes communicative actions of the mouth.[29] Mirror neurons can be further subdivided into "strictly congruent" neurons, for which the performed and observed actions must be identical, and "broadly congruent" ones, for which the two actions need not be identical but must have a similar goal.[30] A significant finding from the standpoint of performance-based neuroaesthetics was that in macaques, the neuron only discharged when the action was actually directed toward a realizable transitive goal and not when the grasping action was carried out without a real object within reach. An experiment designed by Umiltà and colleagues sheds further light on this point. In cases where the experimenter moved his or her arm toward an item concealed by a screen from the monkey, which was aware in advance that an actual object was there, a set of mirror neurons fired that did not fire at all when the monkey knew there was nothing hidden, even though the movement executed was the same in both cases.[31] The issue here is first and foremost that of intention: how minutely does the observer define the goal of an action? Leonardo Fogassi explored this question with a setup in which the monkey watched an experimenter alternately bringing food to his mouth and placing it in a clearly visible container nearby. The mirror neurons coded the action of grasping-to-eat differently from grasping-to-place and were activated in a manner consistent with when the monkey performed the same actions itself.[32] For a goal to be meaningful, it needs to be embedded in the gesture displayed.

The Parma group used neuroimaging to conduct similar tests on humans. These studies confirmed that mirror neurons are present in our brains and are sensorimotor in nature.[33] Our mirror area includes the frontal region of the inferior parietal lobe, the lower region of the precentral gyrus (i.e., the premotor cortex), and the posterior region of the inferior frontal gyrus—in short, the frontal-parietal system. This implies that our responses to the environment do not depend primarily on sensory stimulation, on external inputs to which we react, but instead constitute acts of imagination whereby we simulate internally the psychophysical intention associated with the observed action. We behave as if we were the one acting, based on what is known as a sensorimotor model, which dictates that movement and perception are not separate and consecutive but rather integrated and co-original. Movement is not the displacement of this or that limb but an inclination and a psychosomatic posture conforming (or not) to the correlated surrounding environment: as humans, we move *with* the world. It follows that simulation is neither an automatic reaction (like the grimaces observed in newborn infants forty-two minutes after birth[34]) nor a voluntary and conscious imitation,[35] but the cerebral representation of motor programs without the mandatory production of a concrete and explicit movement. The workings of this *actio in distans* are highly relevant to a performance-based neuroaesthetics for at least two reasons. The first is that the observational response seems indifferent to psychological identity and bodily features. For the macaque, humans and monkeys are equivalent. Is the same true from the human's point of view? Buccino and colleagues recorded the responses of a group of people who were asked to observe a human, a monkey, and a dog engaged in chewing and communicative acts. While the subjects observed the chewing, the brain area associated with ingestive motor acts was activated regardless of which species executed the action. In the case of communicative acts, the posterior region of the inferior frontal gyrus—corresponding to Broca's area—was strongly activated when the human mouthed words (i.e., silent speech), weakly so when the monkey smacked its lips, and not at all so when the dog barked.[36] It would thus appear that we are concerned not with the agent's actual appearance but with its motor identity; what matters is not *who* performs the act but *how* the act is performed.

However, when postulating a body-centered understanding of personal identity, we must be careful to avoid universalizing, as if the body were synonymous with the ancient soul or the modern mind. Otherwise, we would end up replicating the same monist model, rather than deriving the greatest possible epistemic benefit from the change of perspective we are outlining. Another possible outcome of such a universalizing is that the body may be neutralized into a mere instrument for expressing or enacting psychological activity. Privileging the intentional component of the gesture could lead to psychologizing the body's mastery and thinking of it as moved and governed a priori by a principle heterogeneous to it and its spectrum of (in)abilities, when really we encounter it in a plural form, as *bodies* that are made of muscle, bone, blood, water, and skin; have a weight, an age, and a gender; and are first and foremost a spectrum of agencies. However separate it may be from other bodies, it is in contact with them more often than we tend to acknowledge, such as when it makes love, caresses, slaps, breastfeeds, or gestates. Instructive in this regard is Corinne Jola and colleagues' study on dancers, who are specially trained to exploit their sense of agency, not just over their body schema, but also over their kinesthetic abilities. Dancers rely less on vision than nondancers and are better than nondancers at integrating local proprioceptive information from multiple joints to represent the position of their limbs in space. Their bodies are integrated "to form a single coherent body representation to a greater extent than the non-dancers."[37] It follows that "no single 'egocentre' exists for proprioception." The performer is a trained, nurtured, fabricated body, not a naive one. But if this is the case, it is because no naive body exists. Instead of thinking of bodies as *having* expertise, we would do well to adopt the perspective intimated by Paul B. Preciado when he says: "We don't have a body that we come later to reflect upon. We make ourselves a body, we earn our own body. . . . Our bodies and sexualities are collective institutions that we simultaneously inhabit and perform."[38] The binary of cultural/natural, or trained/untrained, is indeed inadequate to explain our material entanglement with a world of images and artifacts. This argument, which the Spanish philosopher articulates on theoretical and even political grounds, has long been at the core of the stage performance. As Étienne Decroux, one of the most talented mimes of the twentieth century, wrote: "[The body] is obliged to dissemble in its movements. Since, unlike the media that the other arts use, it is sensitive, it is bound to suffer in dissembling. If it does not, the body is imitating the body exclusively."[39] In other words, the body is conceived of as a model.

The precise degree of such body making depends on the level of integration of other motor habits into one's own, on a spectrum that runs from assimilating fellow humans' postures to reproducing depictions of them to recording personal movements through motion-capture sensors, projecting their traces on screens, and then reenacting them in order to enhance personal motor skills. According to neuroscientist Marc Jeannerod, simulative activity can indeed respond in itself, without being goal directed. In "The Representing Brain: Neural Correlates of Motor Intention and Imagery" (1994), he posits the existence of a motor imagery of the observed act functionally geared toward motor preparation based on a process of learning by imitation.[40] He further stipulates that the motor imagination, though dependent on observation, is not the same as the dynamic subtype of visual imagination that allows us to imagine scenes of ourselves, other people, or objects moving in space. The piano student looks at the teacher's hands to learn; the teacher looks at the student's hands to check proficiency. For their part, Rizzolatti and Sinigaglia surmise that mirror neurons are linked to the understanding of the motor events of others and thus of others' intentions, based on self-experience: "motor knowledge of our own acts is a necessary and sufficient condition for an immediate understanding of the acts of others."[41] Here, "understanding" refers not to an explanation of the intention but to the ability to recognize the motor event as a "type of action," which, whether occurring on the street or onstage, is central to how choreographer and researcher Ivar Hagendoorn defines dance: "a sequence of movements, not necessarily choreographed, of any length, from two seconds to two hours, whereby the goal of the movement is the movement itself."[42] From this point of view, a dancer is someone who knows a broader or different range of maneuvers compared with a nondancer, though the principle holds for surgeons and typists as well. This concept was foundational for the work of Beatriz Calvo-Merino and colleagues, who recorded brain responses in subjects shown videos of dancers performing classical ballet and capoeira moves. The test subjects fell into three groups: experts in classical ballet, experts in capoeira, and inexpert controls. For experts, observing moves was correlated with greater activation in the premotor cortex, the intraparietal sulcus, the right superior parietal lobe, and the left posterior superior temporal sulcus. The activated areas corresponded to those in which the mirror system is most present. The researchers concluded that "the human brain understands actions by motor simulation," which was related to the subject involved, since it reflected his or her specific motor capabilities. In a subsequent study, they also demonstrated that female participants had a stronger response to the movements of dancers of the same sex than to those of male dancers. As the authors explain, "while all the subjects in our study saw the same actions, the mirror areas of their brains responded quite differently according to whether they could do the actions or not."[43] Therefore, while participants may have had visual familiarity with males and females, what mattered was the common motor practice.[44] The visuomotor system thus apparently responds to patterns of nontransitive or non-goal-oriented actions. If that is so, what is at stake, from the dancer's point of view, is the dynamic among motor habits, personal memory, and expressive and creative freedom—a relationship that, in any case, must be able to solve the problem of "what move to make next."[45] Correspondingly, for the spectator, who occupies a position opposite the dancer, it is a matter of imagining the next move in advance. The result is a bodily dialogue. It would thus seem that the ability to anticipate the near future also depends on mirror neurons, which are—to use Jeannerod's expression—"the vehicle of a contagion."[46] The solidarity between the stage and the auditorium, the merging of performers and viewers, recalls the "spiritual contagion" the church fathers feared to the point of censuring any spectacle that featured the sight of performing and acting bodies. As it happens, in doing so, they were intuitively acknowledging the simulation process and anticipating Brook's assessment of the knowledge implied in theatrical skills.

The second, related reason for interest in the pragmatic dynamics between body/brain systems from the point of view of performance-based neuroaesthetics concerns the integration of actions with the environment. Indeed, the best-advised experimenters know all too well that the main challenge for neurologically testing both artistic production and aesthetic spectatorship lies in matching the fictional/experimental conditions to those of real-life scenarios. Marco Iacoboni and colleagues tried to do this with a special setup. In a first video clip, a hand was shown grasping a cup in a space devoid of meaning (although we might argue that a mere support surface, an objectless background, does not in itself imply an absence so much as a latency of context). This was followed by two scenarios in which no motor act was performed: the first showed a cup on a table set for breakfast, suggesting the act of drinking; the second showed the same cup on a messy table, as it might appear after breakfast, suggesting the acts of moving, clearing the table, and so on. Finally, in a third type of video, a hand was shown grasping the cup in each of the aforementioned scenarios. The observer's neuronal activation was greatest when the act was accompanied by contextual information, especially in the case of drinking. Iacoboni and colleagues concluded that we understand the "the why of an action"—that is, its goal—by inserting it into a chain.[47] These mirror neurons are "*logically* related";[48] they are triggered by a given motor act, such as grasping a cup, but fire during the execution not of the same act but of one functionally related to it, such as bringing to the mouth. What is at stake, from a methodological perspective, is the ecological setting. Pains are taken not to alter the situation in which observation typically unfolds.[49] I consider this effort commendable but pointless: commendable insofar as it is symptomatic of the scientific community's growing awareness of the entanglement of our cognitive and, notably, our spectatorial behaviors; pointless because there is no "original reality" to reproduce. Otherwise, and paradoxical though it may seem, the beholder in a museum would be like a brain in a vat, in line with the well-known thought experiment invoked in chapter 2, which is really a piece of literature. Indeed, the museum would be the natural milieu for the art lover, analogous to the enriched liquid for cerebral matter. In point of fact, every milieu is artifactual and temporary. Neurocognitive experiments, and notably those on mirror neurons, actually extract and typify a fragment of everyday reality, to the point that the respective forms of mediatedness and the apparatuses in which they are embedded overlap not in some arbitrary way but in relation to our embodied (in)abilities. As we have mentioned, the site in which fMRI and PET scans are performed requires the test subject to lie down, immobile, and watch a simplified stimulus via a small monitor. It engineers a situation comparable to that of the theatrical and film spectator, who is silent and still, insulated from the surrounding world by a series of nested frames, from the walls to the proscenium arch to the screen. The test subject's motor activities, like the spectator's, are inhibited, and we might even wonder whether the bourgeois theatrical display did not play some part in shaping this kind of experimental setting. The ecological emphasis, moreover, fuels the idea that by not interfering with the context, we have a chance of capturing the unambiguous data generated by forms and actions. But forms and actions are not there waiting to be discovered any more than bodies are still images in a flipbook. As we have seen, they not only manifest themselves together with the increasingly cyborg gaze that addresses them but are originally contrived and ambiguous, encompassing a plurivocity of intentions to be clarified through success or failure in intersubjective relations. We see someone holding out a cupped hand: is it to get water, or to beg for change? Or is it for a palm reading? Projecting into the body/brain milieu a supposedly objective intention tied to the scientific setting does not bring clarity to the matter. If anything, it amounts to saying that the meaning of the gesture *depends* on the context, when really what constitutes the rationale of an action is motive, means, and opportunity *together with* the apparatus. This can involve the bodily presence of the experimenter. In EEG experiments—that is, in a setting where people are not fully insulated from the environment[50]—he or she even acts or mimes to engage the observer's visuomotor reaction.

Gabriele Sofia, one of the best-informed scholars of the interplay between theater and neuroscience, duly objects that in Umiltà's study on the nonactivation of mirror neurons in a monkey that knew in advance there was nothing behind the screen, the experimenter who executed the "as if" actions lacked the training to make him or her as real within the fiction as an actor would be. Is it indeed "necessary to ponder over the idea of 'mimed' action: is it the copy of a result or the recreation of a process?"[51] To me, the answer is clear from director Eimuntas Nekrosius's instructions to his lead actor playing a jealous Othello in a scene with Desdemona: "if you want the audience to see the strength of your slap, hold up your hand before hitting the face."[52] This question further disrupts the received idea that to be credible a gesture has to be mimetic and reproduce the full design of previous ones like it. Beyond the distinction between transitive and intransitive, gestures may also be classified as either "iconic," where they represent the content of the discourse, such as when we play at imitating someone and thus plan our motions in relation to our observer, or "beat," where they lend a rhythm to the discourse and aid the speaker, regardless of whether he or she is being observed. Du Bos acknowledged these two typologies in his *Critical Reflections on Poetry, Painting and Music*. The body language Du Bos observed in thespians belongs to the iconic category: he termed this "natural," insofar as it was widely attested in the cultures of southern Europe, noting that the actors "[speak] to the eyes."[53] Du Bos characterized rhythmic gestures, as the second type may also be called, as those that serve to punctuate the performance and its beats: "He that makes the gestures, must fall into a cadence at the end of each measure, tho' he be allowed to let some of the time of this measure pass, without gesticulating."[54] The purpose of the gesture here is to set the pace, whose very count has an intentional but nonmimetic core. We may consider, as an example, the violinist's knee that bends on a downbeat and the concert listener's hand that rises slightly in response. (It would be interesting to know how noniconic expressive acts that have no bearing on the rhythm, such as the Neapolitan sneer that sank Wittgenstein's isomorphic theory of language, ought to be classified.)

The distinction between iconic and rhythmic stage gestures supports my view that the representational paradigm is not at all suitable for addressing embodied simulation. If that were so, after grasping the coming action and establishing its underlying intention, we would simply conform based on a model/copy logic. But it is the expressiveness of our bodies in the first place that is contrived, multilayered, and contradictory. To read the mirror system as mimetic, as assenting at each turn to what is presented, would therefore be misleading. Rather, it is highly compatible with a protean conception of human identity and habits. Stafford—echoing Aristotle, who held imitation to be imitation of action and a means of access to knowing,[55] but without taking it as mere duplication—seems to espouse such a view. She argues that the discovery of mirror neurons, conceived of as a kind of wireless communication, restored mimesis to the center of the aesthetic debate after the hiatus brought on by structuralism, and established all cognitive activities as embodied: "If the human instinct for mimesis, imaginative make-believe, or inner rehearsal is a central feature of our cognitive archaeology . . . what more nuanced versions of the self might we now imagine?"[56] I suggest that we come up with a plural understanding of our bodies as the media and the matter of our cognitive performances. It should be noted here that the sensorimotor model is grounded in neurobiological knowledge and took on a theoretical dimension thanks largely to Gallese's project of studying the human as a "symbolic being."[57] He established his theory through a fruitful engagement with both German and French phenomenology. While his initial understanding of inner simulation as an empathetic experience is primarily indebted to Husserl and Lipps, he drew ideas about the cognitive power of the body mainly from Merleau-Ponty, who conceived of it as the organ of thought. Whenever cognition is in play, the body is implicated. This is tantamount to saying that thinking is living and breathing, that speculation takes place in a material environment, and that the world as a multilayered multiplicity exists *for* whoever addresses it, based on varying degrees of presence. The performer responds to other, similar body/brain systems, where each constitutes the pole of a pragmatic—and therefore a cognitive—nexus.[58]

Against the background of such a theoretical undertaking, however, the experiments I have mentioned (which were the first conducted on the visuomotor palette and have thus been quite effective in establishing the broad issues at stake) expose an unthematized circularity between motor identity and the image/imaged loop: the notion of motor identity as presented to us is inherently dependent on the image looked at. This circularity is problematic. The image recalls, revives, and re-creates personal emotional stances embrained in the sensorimotor system, and it does so by way of hints, by testing out possible behaviors. I suggest that, insofar as embodied simulation is construed within the iconic paradigm, its theorization is still bound up with the ocular bias of our culture, and the emphasis on the role of bodily dynamics is in fact largely ancillary. The whole theory indeed finds itself at a crossroads. On the one hand, it accords the figuration of a gesture a decisive role in the processes of motor reproduction by way of simulation. On the other hand, by underscoring the performative aspect of simulation, it reintegrates vision into its original synesthetic dimension. Gallese himself has deemed the term *mirror* inadequate for describing this system, for it suggests a reflection and reproduction of others' actions, as on a two-dimensional screen. It recalls the domain of specular vision, which I myself criticized, by criticizing the mirroring loop whereby visual patterns are supposed refer to the "external world" but are really conceived of as equivalent to it. The image is then absorbed into the viewer. The "as if" clause echoes the model/copy logic, and the notion of a mirror mechanism carries an ocular aftertaste. Therefore, to fully endorse the embodied simulation approach, we need to queer the representational paradigm and inflect our reasoning toward the pairing of effective/potential, which evaluates notions and facts as tempered actions. The visuomotor mechanism provides the organic basis for describing mental and social habits within a framework that is transcendental, insofar as it is material and embedded in apparatuses: the living body as a plexus of situated *virtual agency* is the material transcendental of experience. Andy Clark, who supports the extended perspective, expounds: "humans . . . are integrated but constantly negotiable bodily platforms of sensing, moving, and . . . reasoning. Such platforms extend an open invitation to technologies of human enhancement. They are biologically designed so as to fluidly incorporate new bodily and sensory kits, creating brand new systemic wholes."[59] Some experiments suggest to me that this route is indeed consistent with a *situated* embodied simulation. Umiltà and colleagues set up a now canonical experiment that typifies this thesis. A monkey had to grasp an object, first with normal pliers, which entailed squeezing the grip to close the jaws, and then with reverse pliers (the kind used for eating escargots), which entailed squeezing the grip to open the jaws (in order to surround the object—the snail shell, as it were) and releasing the grip to close them. The result was that "neurons that with normal pliers discharged when the hand was opening, with the reverse pliers discharged when the hand was closing."[60] Conversely, "neurons that with normal pliers discharged when the hand was closing, with the reverse pliers discharged when the hand was opening." This experiment is interesting, not only because it shows that an action's goal outweighs the related motor behavior and its phase-by-phase development prevails over the static pattern, but because it makes clear that if embodied simulation is to be understood as "feeling in," such an immersion entails no assimilation between the observer and the living agent. Any projection into others' gestures enacts distancing, not only from the moving body we look at but from ourselves, for in molding our limbs, we exert agency over them and in turn experience their resistance as well as their compliance. When real simulation occurs, when we physically enact it, we go through the effort of forging our pose, of adapting our respiratory system; we experience our bodily mediality, our inappropriability, our living performativity.

Merleau-Ponty aptly illustrated this dynamic by distinguishing the habitual body of general and prereflective existence from the actual one of active and reflective ownership. The two usually overlap, but in specific situations such as pain their solidarity is patently broken. I submit that whereas pain engenders a clash between the two bodies, at its core every performance stages their distinctness, insofar as it involves two (or more) persons mutually interacting based on a graduated and alternating engagement with their own movements, which, on the one hand, are performed and shown and, on the other, seen and simulated. Actress Daria Lippi and colleagues theorize about the dual nature of the stage action: "Stage acting is not merely acting on a stage, but a kind of acting that is supposed to constantly keep, feed and rule the attention of the spectator. . . . The actor's action is not only directed to an on-stage aim (drinking the glass of water) but also to the audience (stimulating the spectators' attention). The same action has at once two different aims. This gives rise to a 'double intention' of the actor, or better yet, to a *dilated intention* that broadens from the performed action out to the audience."[61] The same goes for spectators who are forced to sit still and remain inactive and yet who feel as if they were moving. I conceive of this dynamic as being chiasmic.[62] This chiasmus is epitomized by the type of performer Brecht associates with Chinese theater, which he viewed in Moscow in 1935 at a performance by Mei Lan-fang's company. The Chinese actor, having donned the mask of the character, will undertake actions such as blatantly sizing up the space available or gauging how to make himself optimally visible to the audience. In this way, he separates the portrayal of the action the character is executing and on which he is intent—as indicated by the focused expression on his face—from the portrayal of the intention that is about to develop and the movement he is going to execute. In Brechtian terms, he separates "mime" from "gesture."[63] Indeed, refraining from a total identification with the character does not mean feigning identification but rather distinguishing between the goal of the character and the path the actor must take to display it. Such a doubleness is likewise required of the critical, nonempathetic spectator whose education the German playwright and director was striving to enhance. The idea that stage performers should *not* migrate into their characters or try to lend them their full bodily agencies has been endorsed by Plessner, who considers the stage actor, independently of acting style, to be an exemplary instantiation of human beings as such, whose positionality is ex-centric: "Man is naturally split, such that he lacks possession of himself. He does not coincide with what he is. . . . His self-presence is based on the gap [between the body he is and the one on which he reflects]."[64] The actor is someone who instantiates and performs this human hiatus, and the interplay between self-presence and self-distancing underwriting the estranged personhood of the actor on stage is also demanded of the spectator: "*As virtual spectators of ourselves and of the world* we must look at the world as though it were a stage."[65] This view of the "as if" reinforces Changeux's claim that spectatorship is a re-creative activity; it is through this re-creation that we "[come] to the point of ascribing 'mental states,' emotions and intentions to the figures that appear in the composition."[66] Not only does art prepare us for life by training us to live, as we have gleaned from Ramachandran and Alexander, but it expounds the inherently dual, self-reflective human nature.

The chiasmic argument I just proposed helps dismantle the image-imaged loop implied in the opinion piece "Motion, Emotion and Empathy in Esthetic Experience," coauthored by Freedberg and Gallese in 2007, which by now represents a classic albeit controversial contribution to visual culture. The authors engage with images as ongoing visuomotor performances from both the maker's and the viewer's points of view. They invoke a heterogenous range of literary and visual sources about the effects of images on beholders, such as Alberti's *On Painting and Sculpture* (1435), which comprises instructions for painting effectively, and various artworks that are claimed to solicit our unlearned gaze for activation. (To be sure, during his lifelong research as an art historian, Freedberg has relied on textual accounts of techniques that provoke vivid imagination through images and, at the same time, on images themselves, gauging their power based on their visual patterns alone. Such an approach, however, reduces the specific agency of each source—from Jesuitic instructions for eliciting co-suffering among believers to paintings of ecstatic figures to depictions of anonymous human bodies in pain—to its empathetic potential.) Freedberg and Gallese surmise that beholding pictures entails a dual relationship between embodied empathetic feelings in the observer and "the representational contents of the works in terms of the actions, objects, emotions and sensations" depicted, on the one hand, and "the quality of the work in terms of the visible traces of the artist's creative gestures," on the other.[67] After rehashing the role of canonical neurons, as well as that of expression, the two authors—drawing on Damasio's work on the "as-if" loop—suggest that the motor act whereby the artist executed the work is in itself what stimulates the beholder's mirror system via the signs on the canvas or on the sculpted material. These marks "are the visible traces of goal-directed movements; hence, they are capable of activating the relevant motor areas in the observer's brain."[68] But the multilayered evidence adduced in this seminal article brings to light two other complicating factors: content and media. As to the first, the lack of historical perspective in the ensemble of visual sources neutralizes any questions of topic and style. The authors grant all artistic production equal standing. They treat Michelangelo's *Slaves*, Caravaggio's *Saint Thomas*, Goya's *Disasters*, Pollock's drippings, and Fontana's cuts interchangeably qua emphatically looked-at visual patterns. Onlookers are therefore aligned with artists, as Damis had been. But whereas Apollonius's beloved pupil described this alignment in terms of a mimetic faculty exerted over the image in its entirety, the authors here present it as a preestablished, preemptive focus on one discrete spot at a time—on a particular area within the framed space or on the molded marble—without regard for any conflicting depictions, light sources, or chromatic nuances in the overall composition or for the contrasting emotions they might convey. The spectatorial performance at play is highly constrained and selective. The counterproposal that this point is ultimately moot because any pattern could elicit a response, and we could therefore focus each time on a different portion of the same image, is not viable, for it merely reinforces the objections I raised: the image is mastered as modular. It would follow from the authors' reasoning that figurativity—be it an abstract landscape or a naturalistic wound, a portrayal of static features or of dynamic ones—does not matter. Yet subsequent experimental studies have presented evidence to the contrary. Notably, the one conducted by Cinzia di Dio and colleagues in 2016 suggests that "the presentation of paintings portraying a human subject produced a stronger activation, compared to nature content stimuli."[69] This may be because we like to recognize a fellow human in the symmetrical silhouette portrayed or because we dislike the feeling of moving through empty space.

My purpose in challenging a neuroaesthetic text by appealing to an experimental study is to drive home the point that any bottom-up analysis of pictures presupposes a figural stimulus (i.e., an iconic content) already seen as a meaningful configuration by the researchers who selected it. Inquiring about the nature of images by way of images themselves opens the door to a meta-analysis of why those specific examples were picked in the first place. When, as a researcher, I choose certain images over others, I make a preliminary selection of the qualities I want my fellow humans to look at. Such a circularity is not problematic per se. As we saw in the previous chapter, artists explore it. Scientists, for their part, neglect it. Assuming we grant my point about the ecological setting, we could hardly expect otherwise. The speciousness lies, rather, in the naive assumption that what neuroscientific research uncovers is given as such, when really, living, breathing scientists are directly implicated in neurological processes and share a comparable cerebral structure with test subjects. A permanent interplay, a continuous feedback loop, is at work among live experimenters, their subjects, the environments of their in vivo inquiries, and the stimuli. Such an interplay is itself chiasmic and cannot be bracketed or thematized as an inner variable of the experimental apparatus. It is rather highlighted through a critical analysis of the material and discursive features implicated in the neuroaesthetic endeavor. I agree with Melissa Littlefield and Jenell Johnson when they contend that a project of this sort, "conceived as a partnership between the arts, sciences, and cultural criticism . . . might ask how art might alter the human brain . . .; how art, politics, and the body intermingle in productive and unexpected ways." This is where performativity and plasticity come together, for "neuroplasticity offers the tantalizing possibility that the brain is also the product of the art it consumes."[70] I do not mean to suggest that cognitive capacities require training or nurturing by the arts, as in a more cultivated version of brain power, nor is it to reclaim the role of culture in a renewed clash with nature. Rather, I am saying that the brain is already a by-product of manufactured stimuli and that to understand their mutual entanglement, we must dismantle the idea of a direct channel between our neurobiological mechanisms and visual and staged artifacts.

The second complicating factor is that Freedberg and Gallese's approach overlooks medium specificity. In his 2007 essay "Empathy, Motion and Emotion," Freedberg discusses the beholding of visual productions ranging from St. John in Giotto's *Lamentation* to the group of terracotta figures in Niccolò dell'Arca's *Lamentation Over the Dead Christ* to Rosso Fiorentino's *Pietà* and Goya's *Disasters of War* to Tyler Hicks's photograph of three grieving Iraqi women and Robert Capa's *Falling Soldier*. He describes the experience of viewing this last piece in a way that could apply to all of them: "One seems to be falling oneself, off-balance, and yet vainly trying to keep oneself upright. It is almost as if, looking at this . . . our bodies respond to it as if *that* body, somehow, were our own. . . . Physical engagement with a picture like this, physical empathy, translates very swiftly into emotion."[71] The question of medium and materiality is rendered nil. Thus, not only are artworks treated interchangeably, but the category of artwork itself is extended to any image made for those who "experience a sense of bodily involvement with the movements that are implied by the physical traces . . . of the creative actions of the producer of the work." From there, what is to stop us from adding Yves Klein's anthropometries, Giuseppe Penone's action sculptures, Anish Kapoor's deposits of earth and pigment, and even my son's play dough creations to the mix? What such a viewer-centric unresponsiveness to media entails is less a cultural indifference than a building from scratch of full-fledged images linked together only by the partial recurrence of analogous patterns. Umiltà and colleagues conducted an experiment on the observation of Fontana's cuts that supports Freedberg and Gallese's working hypothesis, which extends the notion of visual pattern indiscriminately. She used two categories of images as stimuli: three different black-and-white high-resolution digitized images of Fontana's pieces and three high-resolution digitized, graphically modified versions of the same, which displayed the same black-and-white graphic pattern. The researchers found that "the observation of the cuts on canvases activat[ed] the motor representation of the same gesture in the brain of beholders" while the digital images, by contrast, "[were] merely processed as images that are not the manifest consequence of someone's gesture." They remarked, moreover, that "the observation of the static consequences of a motor act is capable of activating its motor representation in the observer's brain."[72] Of course, this study too implies a preliminary selection of visual data, and a highly dubious one at that, since the digitized versions of handmade cuts do not actually present the same information—for instance, concerning depth and shadow—as the originals. However, it is worth mentioning, as it supports the argument that we respond in an analogous motor fashion to the visual landscape, insofar as it is visual, even though we do not respond to every image in exactly the same way. Imaged reality is the only reality, and in the end, the loop is looped.

It was such an expansion of the images' power irrespective of their creation that led Freedberg and Gallese initially to oppose the Kantian notion of "disinterestedness," developed in the *Critique of Judgment*,[73] which dictates that the spectator's focus is distracted from the actual existence of the real thing at stake in the image-making process. In their view, this would imply a certain unfeelingness on the part of the beholder, when really such a disinterestedness in the thing depicted is due to the passionate engagement he or she experiences with the picture. Being involved with an oil painting, a fresco, or a drawing—being affected by its features—is not just fully compatible with being unconcerned with the concrete existence of its (potential) mundane referent; it is actually favored by such a disinterestedness. The neurobiological response associated with beholding unexpectedly upholds a Kantian approach to spectatorship, which stipulates that its target is not the thing but its representation. Engaging via a preconscious simulation heightens the power of the image. The problem arises rather from the fact that in performance-based neuroaesthetics, the image extends in space; its boundaries collapse, and it surrounds us. It is no longer just a portion of a surface in front of me but the full extent of the space in which I can intervene. The visual threshold ceases to be thought of as a frame separating the depiction from the outer world the way a casing separates a window from a wall and establishes a discontinuity between the two; instead, it is an environment between the observer and what is seen. Embodied simulation theory is really a theory of spectatorship that opens toward a narrative of the visual, embodied, and performative power of that spectatorship.

In recent years, Freedberg and Gallese have diverged in their understandings of it. Freedberg, the art historian, has increasingly appealed to a neuro-based theory of beholding. He proposes that the entire aesthetic experience, from absorption to inhibition, self-aware detachment, contemplation, and reflection, "is most clearly understandable, possibly entirely explicable, in neural terms,"[74] to the extent that looking at artworks serves a remedial purpose. In that regard, he has voiced "hopes for therapy via looking,"[75] which is not too far from the views I highlighted in discussing neuroartcriticism. Indeed, starting from an operational notion of artwork as something that retroactively informs our brain,[76] Freedberg helped design an overtly therapeutic study on implied movement with stimuli including pictures of Oskar Schlemmer's sculpture *Abstract Figure* (1923) and Umberto Boccioni's *Unique Forms of Continuity in Space* (1913): "Our results will be relevant in designing more effective, evidence-based programs involving visual art as a potential treatment for patients with both motor and cognitive deficits."[77] The view that art holds medicinal power has been in place since Aristotle expounded his discourse of catharsis. Here it is bound up with the fact that the observation of implied movement in art pieces "can modulate neuroplasticity in fundamental motor circuits."[78]This perspective takes artworks as operating like the virtual reality box in phantom limb syndrome: they intervene in our neurobiological mechanisms. In Freedberg's view, this if anything lends strength to the art historical argument. The nexus of work, maker, and observer, he argues, retrospectively enhances rather than downplays the historical approach, because we cannot know anything about artists of the past "except perhaps by way of their bodily engagements with materials." Artists, it would seem, are anonymous makers whose personal identity is lost to time. All that remains, like a death mask, is the trace of their living body externalized by the gestures they executed: "We still could not see them, only their works. And this would entail an entirely different form of empathy—a representation in ourselves of the movements of the body involved in the labor of producing the work. . . . Though it may seem literalist to some, it actually opens the way to history, not to its denial."[79] The everlasting marks, then, are supposed to safeguard the bodily and intellectual agency of artists.

Such an argument, pious though it may seem, promotes a generalizing notion of human identity, independent of any material-based media history of the agency of works and makers.

The neuroscientist Gallese, meanwhile, reappropriates Alfred Gell's theory of artistic agency whereby both images and artifacts exert humanlike power over human beings and thus qualify as "quasi-persons" based on an intricate web of references and material correspondences.[80] According to Gallese, cultural artifacts are agents that allow us to appreciate the intentionality underwriting their construction. As an example, he points to the spectatorship of Fontana's cuts on canvas "where the consequences of the artist's action—the cuts—lead to the simulation of their causes."[81] However, against the virtually infinite range of humanlike agents that could reveal themselves before our eyes, Gallese takes care to shed light on the night where all cats are black. To maintain a distinction between reality and fiction and to remain faithful to the evolutionistic theory generally implied by neuroscientific knowledge, he has developed the notion of "liberated embodied simulation": "One important context-dependent aspect characterizing our relationship to fictional worlds deals with our distancing from the unrelated external world, which remains at the periphery of our attentional focus, very much like the frame surrounding the image we are beholding. . . . Such distancing . . . liberates new simulative energies. When adopting such aesthetic attitude, our embodied simulation . . . is freed from the burden of modelling our actual presence in daily life."[82] Setting aside the dubious idea that art, by hurting less than real life does, is also less engaging—an idea Changeux, Ramachandran, and Stafford show to be unfounded—such a liberation is, for Gallese, in keeping with the "suspension of disbelief" advocated by Samuel Taylor Coleridge in his *Biographia Literaria* from 1817. In short, Gallese argues that "our relationship with fictional worlds is double edged: on the one hand we pretend them to be true, while, on the other, we are fully aware they are not."[83] As I have stressed over the years, however, proposals like Coleridge's are misleading. The fact is, we do not neatly confer upon images the status of reality despite knowing that they are fictional, just as fictional worlds do not lead us into delusional experiences vis-à-vis rock-solid reality. Rather, we play, with varying degrees of personal agency and social interaction, in multiple arenas—dreams, fiction, and reality—each of which has its own mediatedness, and which are renewed and reenacted by hybridizing elements from one another. Gallese himself has formulated the concept of "reuse" to show us how this occurs. "All simulation type mindreading," he writes, "requires any resemblance of the mental states or processes between the simulator and the target to arise from the reuse of the simulator's own mental states or processes." Therefore, "at bottom it is mental reuse, not resemblance between target and simulator, [that] drive[s] mindreading."[84] Among its epistemic advantages, this proposal divorces embodied simulation from the premise of starting anew at each viewing session. Gallese thereby shows that his research in the area of mirror neurons is not "opaque to the issue of agency" or to the distinction between maker and viewer.[85] It does, however, still conceive of pictures and props, images and gestures, as mere punctual, stand-alone objects, even when it acknowledges their environmental nature and their capacity to alter our sense of near and far, which stems not from metric measurements but from the affective and kinesthetic experience of spatial relations relative to the egocentric body.[86]

After a long detour, we find ourselves again dealing with the problem of the agent and the spectator facing off like two cowboys in a Western film, as we watch and wonder who will shoot first. Indeed, if all kinds of visual artworks, by virtue of being imaged, participate in the same embodying process, if they are nothing but inorganic proxies of living bodies, what, if anything, is the relative agency and the medium specificity of each? There is room to claim that by according a graduated but inexhaustible power to visual patterns, we turn them into nonspecific items endowed with a particular allure, and that neuroaesthetics complements this exercise as a noninvasive technique for displaying the maker's skills and the spectator's cognitive processes. It is art as a low-technology brain imager. Ortega and Vidal offer a similar critique: "Not only [does] neuroaesthetics neglect materiality, and . . . size, texture, color, and scale [but] also does not differentiate between the picture of a hammer and the real hammer one can grasp, the depicted person and the real person with whom one could interact." In other words, "it disregards the radical difference between the 'affordances,' the distinct action possibilities offered by things and by their images."[87] It conceives of artifacts and their representations as a material continuum, an indiscriminate blur. But while I fully support these authors' stance against both the apparent novelty of brainocentrism and its colonizing approach, such an objection applies only if we have already endorsed a face-to-face relation between spectator and maker via the artwork. What if, instead, we were to conceive of spectatorship itself as an authorial performance?

The most significant instantiations of contemporary art production are indeed collaborative undertakings—among artists, performers, sound and light engineers, computer scientists, and institutions—that seek to process the sensorium of voluntary participants based on a shifting hierarchy of embedded human senses, where proximity (touch, sound, smell, motion, balance) is privileged over acuity (sight). These endeavors trigger a questioning of personal resistance to ideological brainocentric narratives. The internationally renowned Danish artist Olafur Eliasson founded the *Institut für Raumexperimente* in Berlin in 2009 and launched a partnership with the Berlin School of Mind and Brain and the Association for Neuroaesthetics. Among his more illustrative pieces—after *The Weather Project* (2003–2004), which featured a never-ending sunset and could be viewed while lying down at the Tate Modern in London—was *Contact* (2014–2015), exhibited at the Foundation Louis Vuitton in Paris. Two straight, mirror-covered walls joined at a right angle, a curved wall emitting a band of monofrequency light, and an upward-sloping floor all combined to manipulate the spatial awareness of the visitor, who had the sense of standing atop a giant sphere.[88] James Turrell's *Roden Crater*, located in the Painted Desert region of northern Arizona and finally set to open after decades of planning, is conceived of as an immersive installation inside a volcanic cinder cone, far from "manmade distractions"[89]—that is, from mere artifacts. With experiences such as *Remote X*, the Swiss theater company Rimini Protokoll sets up remote-controlled walks through historic cities in which participants, guided by prerecorded voices, become at once first-person agents and spectators of others' actions in streets, churches, and other public places.[90] Theater director and visual artist Romeo Castellucci, together with composer Scott Gibson, staged *Il Terzo Reich* (The Third Reich) (2021) in which thousands of words were projected one after the other onto a huge screen at a speed surpassing our retinal and mnemonic capacity to retain images; each projection was accompanied by a harsh audio input like an explosion. A new type of synesthetic *Gesamtkunstwerk*, a total artwork in which the artifact coalesces with the experience we have of it, is at hand. Beholding itself takes on the features of a performance—a perceptual, mnemonic, emotional performance. The following quote from a 2017 *New York Times* review of choreographer Jody Oberfelder's interactive project *The Brain Piece* is enlightening: "Ms. Oberfelder said that she hoped they [the viewers] remain 'alive and active in their seats.' As she put it, 'I want people to perceive dance like smell, like something that goes right to your brain.'"[91] This aspiration is corroborated by more authoritative sources as well. Johannes Volkelt, more than a century earlier, stressed the degrees of living movement whereby the spectator observing the staged gesture is transposed: "It is very common for the subject engaged in aesthetic contemplation to take part in the observed movement by way of hints. Thus, a theater spectator can take part with a slight hint of a rigid posture or a rhythmic dance."[92] Eugenio Barba observes that even when spectators watch a performance in a language they do not know, they follow its rhythm and "dance although seated."[93] Freedberg, who is convinced we can experience the emotional essence of a painting "even without the most basic knowledge of the elements of the story,"[94] would certainly agree. To the beholder's motor imagination, visual art represents the zero degree of the stage action, and theater goes back to being a *tableau vivant*. There is thus no *work* of art, no reason to think that art comes *after* making.

At the end of the day, the medial awareness and the aesthetic and neurobiological thought involved could wind up being propelled by the inertial force of the ocular epistemic paradigm. In that case, no doubt, they would reinforce received dualistic ideas. Both neuro-based art and art-based neuroscience would then mimic the long-standing notion of the solitary, unmovable, cognitive subject who wields control over external processes but is inherently vulnerable to the surrounding environment that stands in the way of peace and silence. The result would be much ado about nothing. Alternatively, such endeavors may turn our living bodies, in their entanglement with media and materials, into the very site of art production, and the corresponding knowledge could be construed and deployed toward a more analytical understanding of those situated, dual, chiasmic, profoundly historical, and mortal humans that we are. Indeed, as that living, performing being, we owe our very intelligence and survival to the (im)possibility of a continuous exchange with the world in which we are immersed and with which we negotiate our inherent imbalance. In the future, we may marvel at that abstract and autonomous mind we know we are not for presuming to live by itself, optimally unburdened by interactions with others' bodies, feelings, thoughts, and artifacts. We may indeed look back at the disembodied Self as the greatest fiction to behold.

# NOTES

## *NEUROAESTHETICS* RELOADED

1. Marcos Nadal and Marcus T. Pearce, "The Copenhagen Neuroaesthetics Conference: Prospects and Pitfalls for an Emerging Field," *Brain and Cognition* 76 (2011): 174.

## 1. 1994: PUTTING NEUROAESTHETICS ON THE MAP

1. Joseph R. Hixson, "New Seeing-Eye Machines . . . Look Inside Your Body, Can Save Your Life," *Vogue*, July 1983, 238–39, 254–56. As anthropologist Joseph Dumit remarks: "The article does not need to be read to be understood. The juxtaposition of words and images brings home quite forcefully that the three colored ovals are brain scans, and that the three scanned brains are different. Presumably, these brains belong to different people, three different kinds of persons because their brains are not the same. . . . These images persuade viewers to equate person with brain, brain with scan, and scan with diagnosis." Joseph Dumit, "Is It Me or My Brain? Depression and Neuroscientific Facts," *Journal of Medical Humanities* 24, no. 1–2 (2003): 36.
2. Giovanni Frazzetto and Suzanne Anker, "Neuroculture," *Perspectives* 10 (2009): 816. See also Giovanni Frazzetto, "Neural Networking in Manhattan," *Nature* 451 (February 14, 2008): https://doi.org/10.1038/451772a.
3. Nima Bassiri, "Who Are We, Then, If We Are Indeed Our Brains? Reconsidering a Critical Approach to Neuroscience," in *Neuroscience and Critique: Exploring the Limits of the Neurological Turn*, ed. Jan De Vos and Ed Pluth (New York: Routledge, 2016), 44.
4. American Psychological Association, *Functional Magnetic Resonance Imaging: A New Research Tool* (Washington, DC: American Psychological Association, 2007), accessed July 14, 2021, https://www.yumpu.com/en/document/read/10408039/functional-magnetic-resonance-imaging-a-new-research-tool.

5. Frazzetto and Anker, "Neuroculture," 815.
6. On the media coverage of neurotechnologies and its impact on both society and science, see Garret O'Connell, Janet De Wilde, Jane Haley, Kirsten Shuler, Burkhard Schafer, Peter Sandercock, and Joanna M. Wardlaw, "The Brain, the Science and the Media," *EMBO Reports* 12, no. 7 (2011): 630–36.
7. Jan De Vos and Ed Pluth, introduction to *Neuroscience and Critique*, 2.
8. Vittorio Gallese, "Mirror Neurons and Art," in *Art and the Senses*, ed. Francesca Bacci and David Melcher (Oxford: Oxford University Press, 2013), 461.
9. Bree Chancellor and Anjan Chatterjee, "Brain Branding: When Neuroscience and Commerce Collide," *AJOB Neuroscience* 2, no. 4 (2011): 24.
10. Fernando Vidal, "Brainhood, Anthropological Figure of Modernity," *History of the Human Sciences* 22, no. 1 (2009): 10.
11. Alain Ehrenberg, "Le Cerveau 'social': Chimère épistémologique et vérité sociologique," *Esprit*, January 2008, 101.
12. Victoria Pitts-Taylor, *The Brain's Body: Neuroscience and Corporeal Politics* (Durham, NC: Duke University Press, 2016), 21.
13. This is harshly criticized in Raymond Tallis, *Aping Mankind: Neuromania, Darwinitis and the Misrepresentation of Humanity* (Durham, NC: Acumen, 2011).
14. Chancellor and Chatterjee, "Brain Branding," 23–24.
15. Brain Bullet, accessed July 20, 2021, https://www.brainbullet.com/index.html.
16. Warren Neidich, *The Glossary of Cognitive Activism (for a Not So Distant Future)* (Berlin: Archive Books, 2019).
17. Chancellor and Chatterjee, "Brain Branding," 23–24.
18. Micaela Marini Higgs, "How to Get the Most Out of Art (Even When You're Not Sure You Get It)," New York Times, August 4, 2018, https://www.nytimes.com/2018/08/14/smarter-living/how-to-get-the-most-out-of-art-even-when-youre-not-sure-you-get-it.html.
19. For the rhetorical nature of the contemporary brain, see Davi Johnson Thornton, *Brain Culture: Neuroscience and Popular Media* (New Brunswick, NJ: Rutgers University Press, 2011).
20. Luis-Alfonso Arráez-Aybar, Pedro Navia-Álvarez, Talia Fuentes-Redondo, and José-L. Bueno-López, "Thomas Willis, a Pioneer in Translational Research in Anatomy (on the 350th Anniversary of *Cerebri Anatome*)," *Journal of Anatomy* 226, no. 3 (2015): 289–300.
21. Andrew P. Wickens, *A History of the Brain: From Stone Age Surgery to Modern Neuroscience* (New York: Psychology Press, 2015), 58.
22. An effective critique of Kandel's claim, and more broadly of the two-culture issue, can be found in Michael Kubovy, "Neuroaesthetics: Maladies and Remedies," *Art & Perception* 8 (2020): 1–26.
23. Chris McManus, "Measuring the Culture of C. P. Snow's *Two Cultures*," *Empirical Studies of the Arts* 24, no. 2 (2006): 219–27.
24. William Hirstein, "Neuroaesthetics: Responding to the Critics," *Psychology Today*, December 18, 2012.

25. Thomas J. Carew and Ramaswami Mani, "The Neurohumanities: An Emerging Partnership for Exploring the Human Experience," *Neuron* 108 (2020): 590–93.
26. Nicolas J. Bullot, William P. Seeley, and Stephen Davies, "Art and Science: A Philosophical Sketch of Their Historical Complexity and Codependence," *Journal of Aesthetics and Art Criticism* 75, no. 4 (2017): 453–63.
27. James Croft, "The Challenges of Interdisciplinary Epistemology in Neuroaesthetics," *Mind, Brain and Education* 5, no. 1 (2011): 6.
28. Raymond Tallis, "The Limitations of a Neurological Approach to Art," *Lancet* 372 (2008): 19.
29. Manuel Anglada-Tort and Martin Skov, "What Counts as Aesthetics in Science? A Bibliometric Analysis and Visualization of the Scientific Literature from 1970 to 2018," *Psychology of Aesthetics, Creativity, and the Arts* (2020): https://doi.org/10.1037/aca0000350.
30. Roland Barthes, "From Work to Text," trans. Richard Howard, in *The Rustle of Language* (Berkeley: University of California Press, [1971] 1989), 56.
31. Fernando Vidal, "La neuroesthétique, un esthétisme scientiste," *Revue d'Histoire des Sciences Humaines* 25, no. 2 (2011): 251.
32. Neuroaesthetics has, from the outset, excited interest and drawn funding from scholars, institutions, foundations, and research centers. In spring 1995, the Center Leo Apostel and Brussels Free University hosted "Einstein Meets Magritte: An Interdisciplinary Reflection in Science, Nature, Art, Human Action and Society," organized on the premise that both art and science start from an original level of experience. In 2002, the Getty Museum held a conference, "Frames of Viewing: The Brain, Cognition, and Art," at its venue in Brentwood, California, in collaboration with the Center for Advanced Study in the Behavioral Sciences (April 3–6) with, among others, Aage Brandt, Martin Kemp, George Lakoff, Steven Pinker, Vilayanur S. Ramachandran, Barbara Maria Stafford, Mark Turner, and Semir Zeki; see Jonathan Miller, "The Gaze: Looking as It Appears in Pictures," Getty Research Institute, accessed July 20, 2021, https://www.getty.edu/research/exhibitions_events/events/miller.html. In spring 2007, the European Union's European Science Foundation (ESF) program included neuroaesthetics in its Young Researchers Forum "Disciplines and Borders: Humanities Research in an Age of Interdisciplinarity"; see European Science Foundation, "Disciplines and Borders: Humanities Research in an Age of Interdisciplinarity," accessed July 2021, http://archives.esf.org/fileadmin/Public_documents/Publications/SCH%20Manifesto.pdf. The curators of the exposition "Monumenta 2008," held at the Grand Palais in Paris and dedicated to Richard Serra, called on neurologist Jean-Pierre Changeux—who had taught courses in neuroaesthetics at the Collège de France from 2003 until he retired from active teaching in 2006—to discuss the American sculptor's work *Promenade*. He analyzed Serra's enormous walls from the standpoint of subjectivity and the spectator's neural response. In 2013, a major institutional step was achieved with the establishment of the Max Planck Institute for Empirical Aesthetics in Frankfurt. Although this institute does not identify itself with neuroaesthetics, it includes a neuroscience department, and its self-description as an interdisciplinary enterprise that "focuses primarily on the foundations of aesthetic

evaluation, perception and experience" is broadly in line with the project. In 2020, the Fondazione Prada launched the series "Human Brains"; see Fondazione Prada, "Human Brains," accessed July 20, 2021, https://humanbrains.fondazioneprada.org/it/.

33. Mazda Adli, Maximilian Berger, Eva-Lotta Brakemeier, Ludwig Engel, Joerg Fingerhut, Ana Gomez-Carillo, Rainer Hehl, et al., "Neurourbanistik—ein methodischer Schulterschluss zwischen Stadtplanung und Neurowissenschaften," *Die Psychiatrie* 13, no. 2 (2016): 70–78; Harry Francis Mallgrave, *Architecture and Embodiment: The Implications of the New Sciences and Humanities for Design* (London: Routledge, 2013).
34. Nicolai Rostrup, "Environmental Neuroaesthetics: Landscapes, Places and Architecture," in *An Introduction to Neuroaesthetics: The Neuroscientific Approach to Aesthetic Experience, Artistic Creativity, and Arts Appreciation*, ed. Jon O. Lauring (Copenhagen: Museum Tusculanum, 2014), 185–210.
35. Anjan Chatterjee, "Neuroaesthetics," in *The Oxford Encyclopedia of Aesthetics*, ed. Michael Kelly, 2nd ed. (New York: Oxford University Press, 2014), 481–91.
36. Giulio Preti, *Retorica e logica* (Turin: Einaudi, 1968), 14.
37. James Elkins, *Six Stories from the End of Representation: Images in Painting, Photography, Astronomy, Microscopy, Particle Physics and Quantum Mechanics, 1980–2000* (Stanford, CA: Stanford University Press, 2008), 8.
38. Jean-Pierre Changeux and Paul Ricoeur, *What Makes Us Think? A Neuroscientist and a Philosopher Argue About Ethics, Human Nature and the Brain*, trans. M. B. DeBevoise (Princeton, NJ: Princeton University Press, [1998] 2000), 24–25.
39. Philippe Descola, *Beyond Nature and Culture* trans. Janet Lloyd, (Chicago: University of Chicago Press, [2005] 2014).
40. Louise Barrett, "The Evolution of Cognition: A 4E Perspective," in *The Oxford Handbook of 4E Cognition*, ed. Albert Newen, Leon De Bruin, and Shaun Gallagher (Oxford: Oxford University Press, 2018), 721.
41. Wilhelm Dilthey, *Selected Works Volume I: Introduction to the Human Sciences*, ed. Rudolf A. Makkreel and Frithjof Rodi (Princeton, NJ: Princeton University Press, [1883] 1989), 61.
42. Edmund Husserl, *The Crisis of European Sciences and Transcendental Phenomenology: An Introduction to Phenomenological Philosophy*, trans. David Carr (Evanston, IL: Northwestern University Press, [1936] 1970).
43. Erwin Panofsky, *Galileo as a Critic of the Arts* (The Hague: Springer, [1954] 2014).
44. Frank J. Sulloway, *Freud, Biologist of the Mind: Beyond the Psychoanalytic Legend* (Cambridge, MA: Harvard University Press, [1979] 1982).
45. Eric R. Kandel, *The Age of Insight: The Quest to Understand the Unconscious in Art, Mind, and Brain, from Vienna 1900 to the Present* (New York: Random House, 2012), 45.
46. Alois Riegl, *Late Roman Art Industry*, trans. Rolf Winkes (Rome: Giorgio Bretschneider, [1901] 1985).
47. Paul K. Feyerabend, *Wissenschaft als Kunst* (Frankfurt am Main: Suhrkamp, 1984).
48. Antonio R. Damasio, *Descartes' Error: Emotion, Reason, and the Human Brain* (New York: Avon, 1994), 172; Immanuel Kant, *Critique of Pure Reason*, trans. and ed. Paul S. Guyer and Allen W. Wood (Cambridge: Cambridge University Press, [1781] 1998).

49. Antonio R. Damasio, *Descartes' Error: Emotion, Reason, and the Human Brain*, reprint ed. (New York: Penguin, [1994] 2005), xiv.
50. Dilthey, *Selected Works Volume I*, 67.
51. Pitts-Taylor, *The Brain's Body*, 145.
52. John Brockman, *The Third Culture: Beyond the Scientific Revolution* (New York: Simon & Schuster, 1995).
53. Cliodhna O'Connor and Helene Joffe, "How Has Neuroscience Affected Lay Understandings of Personhood? A Review of the Evidence," *Public Understanding of Science* 22, no. 3 (2013): 256.
54. Francesco Ortega and Fernando Vidal, *Being Brains: Making the Cerebral Subject* (New York: Fordham University Press, 2017).
55. Vidal, "Brainhood, Anthropological Figure of Modernity," 7.
56. Nikolas Rose and Joelle M. Abi-Rached, *Neuro: The New Brain Sciences and the Management of the Mind* (Princeton, NJ: Princeton University Press, 2013), 220.
57. For a clear and concise critique of "affective brainocentrism," see Giovanna Colombetti and Eder Zavala, "Are Emotional States Based in the Brain? A Critique of Affective Brainocentrism from a Psychological Perspective," *Biology and Philosophy* 34, no. 45 (2019): https://doi.org/10.1007/s10539-019-9699-6.
58. O'Connor and Joffe, "How Has Neuroscience Affected Lay Understandings of Personhood?," 255.
59. Catherine Malabou, *Ontology of the Accident: An Essay on Destructive Plasticity*, trans. C. Shread (Cambridge: Polity, [2009] 2012), 25.
60. John Locke, *An Essay Concerning Human Understanding*, ed. Pauline Phemister (Oxford: Oxford University Press, [1690] 2008), bk. 2, chap. 3, sec. 1.
61. Eric R. Kandel, quoted in Ellen Switzer, "Brain Work," *Vogue*, November 1983, 468.
62. One film tells this story particularly well. In 1994, the BBC produced Kevin Hull's documentary *Einstein's Brain* about Kenji Sugimoto, a Japanese Einstein worshipper on a quest to obtain a slice of his hero's brain. Sugimoto's story is recounted from his point of view as he travels around the United States in search of the organ. Einstein's brain, removed from his skull during his autopsy, has since been handled, photographed, sliced on a kitchen cutting board, stored in jam jars, and distributed to scholars and fans. Sugimoto's dedication is ultimately rewarded when he receives a piece of the brain, specially cut to his satisfaction.
63. Jean-Pierre Changeux, *The Good, the True and the Beautiful: A Neuronal Approach*, trans. and rev. Laurence Garey (New Haven, CT: Yale University Press, [2008] 2012), 5. For the conference, see Minerva Foundation, "2002 Conference on Neuroaesthetics: The Pleasure of Art as Sensed by the Brain," accessed July 20, 2021, https://www.minervaberkeley.org/cf2002.
64. Camilo J. Cela-Conde and Francisco J. Ayala, "Art and Brain Co-Evolution," in *Art, Aesthetics and the Brain*, ed. Joseph P. Huston, Marcos Nadal, Francisco Mora, Luigi F. Agnati, and Camilo J. Cela-Conde (Oxford: Oxford University Press, 2015), 408–25.
65. See Oshin Vartanian and Vinod Goel, "Neuroanatomical Correlates of Aesthetic Preferences for Paintings," *Neuroreport* 15 (2004): 893–97; Hideaki Kawabata and Semir Zeki,

"The Neural Correlates of Beauty," *Journal of Neurophysiology* 91 (2004): 1699–705; Camilo J. Cela-Conde, Gisèle Marty, Fernando Maestú, Tomás Ortiz, Enric Munar, Alberto Fernández, et al., "Activation of the Prefrontal Cortex in the Human Visual Aesthetic Perception," *Proceedings of the National Academy of Sciences* 101 (2004): 6321–25.

66. Claus-Christian Carbon, "Empirical Aesthetics: In Quest of a Clear Terminology and Valid Methodology," in *Exploring Transdisciplinarity in Art and Sciences*, ed. Zoï Kapoula, Emmanuelle Volle, Julien Renoult, and Moreno Andreatta (Cham, Switzerland: Springer Verlag, 2018), 117.
67. Antonio R. Damasio and Hanna Damasio, "Infinis d'Asie," in *Catalog of the Exhibition of Jean-Baptiste Huynh* (Paris: Museum of Asian Arts, Guimet National Museum of Asian Arts, 2019); Antonio R. Damasio, "The Photographs of Janet Sternburg," in *Fisher Museum Exhibition Catalogue* (Los Angeles: University of Southern California, 2018); Antonio R. Damasio, "Film, Mind, and Emotion: The Brain Perspective: Movies on the Mind," in *Psychology and Film Since Sigmund Freud* (Berlin: Deutsche Kinemathek, Museum für Film und Fernsehen, 2006), exhibition catalogue; Antonio R. Damasio, "Euthanasia Coaster," in *Design and Violence*, ed. Paola Antonelli and Jamer Hunt (New York: Museum of Modern Art, 2015), 194–97, accessed July 21, 2020, https://www.moma.org/interactives/exhibitions/2013/designandviolence/euthanasia-coaster-julijonas-urbonas/, exhibition catalogue; Antonio R. Damasio, "On Seeing Things and the Power of Photography," in *See the Light*, ed. Brit Salvesen (Los Angeles: Marjorie and Leonard Vernon Collection, Los Angeles County Museum of Art, 2013), exhibition catalogue.
68. Antonio R. Damasio, "L'omeostasi artistica," in "Van Gogh e i neuroni: L'arte come esperienza cognitiva," ed. Gloria Origgi, special issue, *MicroMega* 2 (2007): 186.
69. See, for example, Kay Young, *Imagining Minds: The Neuro-Aesthetics of Austen, Eliot, and Hardy* (Columbus: Ohio State University Press, 2010).
70. Eric R. Kandel, *Reductionism in Art and Brain Science: Bridging the Two Cultures* (New York: Columbia University Press, 2016), 199.
71. John Hyman, "Art and Neuroscience," in *Beyond Mimesis and Convention: Representation in Art and Science*, ed. Roman Frigg and Matthew C. Hunter (Heidelberg: Springer, 2010), 245.
72. Semir Zeki, *Inner Vision: An Exploration of Art and the Brain* (Oxford: Oxford University Press, 1999), 2.
73. Semir Zeki and Tomohiro Ishizu, "The 'Visual Shock' of Francis Bacon: An Essay in Neuroesthetics," in *Bacon and the Mind: Art, Neuroscience and Psychology*, ed. Martin Harrison (London: Estate of Francis Bacon, 2019), 121.
74. Lamberto Maffei and Adriana Fiorentini, *Arte e cervello* (Bologna: Zanichelli, 1995), viii.
75. Vilayanur S. Ramachandran and William Hirstein, "The Science of Art: A Neurological Theory of Aesthetic Experience," *Journal of Consciousness Studies* 6, no. 6–7 (1999): 15–51.
76. Vilayanur S. Ramachandran, *The Artful Brain*, Forum for the Future of Higher Education, accessed July 6, 2021, http://forum.mit.edu/articles/the-artful-brain/.

77. Jean-Pierre Changeux, "Art and Neuroscience," *Leonardo* 27, no. 3 (1994): 189.
78. For a study on the extent to which the two disagreed in their approaches to the psychology of art, see Ian Verstegen, "Revisiting Arnheim and Gombrich in Social Scientific Perspective," *Journal of Aesthetics and Art Criticism* 76, no. 1 (Winter 2018): 45–55.
79. Ernst Hans Gombrich, *Art and Illusion: A Study in the Psychology of Pictorial Representation* (Princeton, NJ: Princeton University Press, 1960).
80. Christopher S. Wood, "E. H. Gombrich's 'Art and Illusion: A Study in the Psychology of Pictorial Representation' " [1960], *Burlington Magazine* 151, no. 1281 (December 2009): 836–39.
81. Ricciotto Canuto, "L'Art Cérébriste," *Le Figaro*, February 9, 1914, 1–2. Translated by Samuel Fleck.
82. Gilles Deleuze, *Cinema 2: The Time-Image*, trans. Hugh Tomlinson and Robert Galeta (Minneapolis: University of Minnesota Press [1985] 1997), 211.
83. Arthur C. Danto, "The Artworld," *Journal of Philosophy* 61, no. 19 (1964): 571–84. Danto's theoretical analysis is so well suited to the neuroaesthetic approach that it is used on a practical level. See, for example, Helmut Leder, Benno Belke, Andries Oeberst, and Dorothee Augustin, "A Model of Aesthetic Appreciation and Aesthetic Judgments," *British Journal of Psychology* 95 (2004): 493. The authors state: "A work of art is the input for the model. Aesthetic experiences often require a pre-classification of an object as art. This pre-classification can be assured by a number of possible context features. The appearance of an object in an art exhibition, in a museum or art gallery is a strong contextual cue for classifying an object as one that warrants aesthetic processing."
84. Arthur C. Danto, "Seeing and Showing," *Journal of Aesthetics and Art Criticism* 59, no. 1 (2001): 9.
85. Alva Noë, "Art and the Limits of Neuroscience," *New York Times*, December 4, 2011, https://opinionator.blogs.nytimes.com/2011/12/04/art-and-the-limits-of-neuroscience/.
86. Martin Skov and Marco Nadal, "A Farewell to Art: Aesthetics as a Topic in Psychology and Neuroscience," *Perspectives on Psychological Science* 15, no. 3 (2020): 630, 636.
87. Theodor Lipps, *Ästhetik: Psychologie des Schönen und der Kunst* (Hamburg: L. Voss, 1903–1906), 2:1.
88. Michael Meulders, *Helmholtz: From Enlightenment to Neuroscience*, trans. Laurence Garey (Cambridge, MA: MIT Press, [2001] 2012).
89. August Schmarsow, "Gemeinschaft der Sinnesgebiete im schöpferischen Akt," *Zeitschrift für Ästhetik und allgemeine Kunstwissenschaft* 24 (1930): 2–3.
90. Marc Jimenez, *Art et technosciences: Bioart et neuroesthétique* (Paris: Klincksieck, 2016), 100.
91. Marcos Nadal, Antoni Gomila, and Alejandro Gàlvez-Pol, "A History for Neuroaesthetics," in *An Introduction to Neuroaesthetics: The Neuroscientific Approach to Aesthetic Experience, Artistic Creativity, and Arts Appreciation*, ed. Jon O. Lauring (Copenhagen: Museum Tusculanum, 2014), 3–49.

92. Nadal, Gomila, and Gàlvez-Pol, "A History for Neuroaesthetics," 17.
93. Gernot Gerger and Helmut Leder, "Titles Change the Esthetic Appreciation of Paintings," *Frontiers in Human Neuroscience* 9, no. 464 (2015): https://doi.org/10.3389/fnhum.2015.00464.
94. Dahlia W. Zaidel, "Hemispheric Asymmetries in Memory for Incongruous Scenes," *Cortex* 24 (1988): 231–44.
95. Marilyn Freimuth and Seymour Wapner, "The Influence of Lateral Organization in the Evaluation of Paintings," *British Journal of Psychology* 70 (1979): 211–18; John P. McLaughlin, Pamela Dean, and Philip Stanley, "Aesthetic Preference in Dextrals and Sinistrals," *Neuropsychologia* 21 (1983): 147–53.
96. Maffei and Fiorentini, *Arte e cervello*, 68.
97. Ernst Cassirer, "Nature-Concepts and Culture-Concepts," in *The Logic of the Humanities*, trans. Clarence Smith Howe (New Haven, CT: Yale University Press, 1961), 118.
98. Gregory Moore, "Art and Evolution: Nietzsche's Physiological Aesthetics," *British Journal for the History of Philosophy* 10, no. 1 (2002): 109–26.
99. Miriam Bratu Hansen, *Cinema and Experience: Siegfried Kracauer, Walter Benjamin, and Theodor W. Adorno* (Berkeley: University of California Press, 2012).
100. Jean-Baptiste Du Bos, *Critical Reflections on Poetry, Painting and Music*, ed. James O. Young and Margaret Cameron (London: Brill, [1719] 2021), 2.
101. Edmund Burke, *A Philosophical Enquiry Into the Origin of Our Ideas of the Sublime and Beautiful*, ed. Paul Guyer (Oxford: Oxford University Press, [1757] 1990), sec. 4, pt. 13, 113–14.
102. Nadal and Pearce, "The Copenhagen Neuroaesthetics Conference," 172.
103. Anjan Chatterjee and Oshin Vartanian, "Neuroaesthetics," *Trends in Cognitive Sciences* 18, no. 7 (2014): 370.
104. Marcus T. Pearce, Dahlia W. Zaidel, Oshin Vartanian, Martin Skov, Helmut Leder, Anjan Chatterjee, and Marcos Nadal, "Neuroaesthetics: The Cognitive Neuroscience of Aesthetic Experience," *Perspectives on Psychological Science* 11 (2016): 266, 270.
105. Anjan Chatterjee and Oshin Vartanian, "Neuroscience of Aesthetics," *Annals of the New York Academy of Sciences* 1369, no. 1 (2016): 189.
106. Chatterjee and Vartanian, "Neuroaesthetics," 370.
107. Chatterjee and Vartanian, "Neuroscience of Aesthetics," 189.
108. Beatriz Calvo-Merino and Patrick Haggard, "Neuroaesthetics of Performing Arts," in *Art and the Senses*, ed. Francesca Bacci and David Melcher (New York: Oxford University Press, 2011): 529–41.
109. Martin Skov and Oshin Vartanian, "What Have We Learned Since *Neuroaesthetics*?," preface to the second edition of the Korean translation of *Neuroaesthetics* (n.p.: Korea Books Medio, 2019).
110. Pearce et al., "Neuroaesthetics," 266, 270.
111. Helmut Leder and Pablo P. L. Tinio, "Experimental Aesthetics," in *An Introduction to Neuroaesthetics: The Neuroscientific Approach to Aesthetic Experience, Artistic Creativity, and Arts Appreciation*, ed. Jon O. Lauring (Copenhagen: Museum Tusculanum, 2014), 64.

## 2. NEUROAESTHETICS: CEREBRAL ATTRIBUTES AND BODILY GHOSTS

1. Noel Carroll, Margaret Moore, and William P. Seeley, "The Philosophy of Art and Aesthetics, Psychology and Neuroscience: Studies in Literature, Music and Visual Art," in *Aesthetics Science: Connecting Minds, Brains, and Experience*, ed. Arthur P. Shimamura and Stephen E. Palmer (New York: Oxford University Press, 2012), 48.
2. Steven Brown, "Toward a Unification of the Arts," *Frontiers in Psychology* 9 (2018): https://doi.org/10.3389/fpsyg.2018.01938.
3. Martin Skov and Marco Nadal, "Art Is Not Special: An Assault on the Last Lines of Defense Against the Naturalization of the Human Mind," *Reviews in the Neurosciences* 29, no. 6 (2018): 701.
4. Anjan Chatterjee, "Neuroaesthetics: A Coming of Age Story," *Journal of Cognitive Neuroscience* 23, no. 1 (2011): 54. For a critique of the parallelism, see Peer F. Bundgaard, "Feeling, Meaning, and Intentionality—A Critique of the Neuroaesthetics of Beauty," *Phenomenology and the Cognitive Sciences* 14, (2015): 781–801.
5. Norman Bryson, "Natural Attitude," in *Vision and Painting: The Logic of the Gaze* (New Haven, CT: Yale University Press, 1983), 1–12.
6. Anjan Chatterjee, "Neuroaesthetics: Growing Pains of a New Discipline," in Shimamura and Palmer, *Aesthetic Science*, 299.
7. Anjan Chatterjee, Page Widick, Rebecca Sternschein, William B. Smith, and Bianca Bromberger, "The Assessment of Art Attributes," *Empirical Studies of the Arts* 28, no. 2 (2010): 219.
8. Jane E. Else, Jason Ellis, and Elizabeth Orme, "Art Expertise Modulates the Emotional Response to Modern Art, Especially Abstract: An ERP Investigation," *Frontiers in Human Neuroscience* 9 (2015): https://doi.org/10.3389/fnhum.2015.00525.
9. Semir Zeki, "Improbable Areas in Color Vision," in *The Visual Neurosciences*, ed. Leo M. Chalupa and John Simon Werner (Cambridge, MA: MIT Press, 2004), 2:1029.
10. Alexander Gottlieb Baumgarten, *Aesthetica* (Hamburg: F. Meiner, [1750] 2007), sec. 1.
11. Andrea Pinotti, "Neuroestetica, estetica psicologica, estetica fenomenologica: le ragioni di un dialogo," *Rivista di estetica* 37, no. 1 (2008): 147–68.
12. Rudolf Arnheim, *Visual Thinking* (Berkeley: University of California Press, 1969), 134.
13. Mark Rollins, "What Monet Meant: Intention and Attention in Understanding Art," in "Art, Mind and Cognitive Science," ed. Jerrold Levinson and Jenefer Robinson, special issue, *Journal of Aesthetics and Art Criticism* 62, no. 2 (Spring 2004): 175.
14. Luigi Francesco Agnati, Achille Agnati, Francisco Mora, and Kjell Fuxe, "Does the Human Brain Have Unique Genetically Determined Networks Coding Logical and Ethical Principles and Aesthetics? From Plato to Novel Mirror Networks," *Brain Research Reviews* 55 (2007): 74.
15. George K. York, "Localization in the Visual Brain," in *The Neurobiology of Painting*, ed. Frank Clifford Rose (San Diego: Elsevier, 2006), 58.
16. Wassily Kandinsky, *Concerning the Spiritual in Art*, trans. M. T. H. Sadler (New York: Dover, [1912] 1977), 25.

17. Leonardo Da Vinci, *Libro di pittura*, facsimile edition of the Codex Urbinas Vaticanus Latinus 1270, ed. Carlo Pedretti (Florence: Giunti, 1995), pt. 1, sec. 3, 132–33. Translation in Claire Farago, *Leonardo Da Vinci's Paragone: A Critical Interpretation with a New Edition of the Text in the Codex Urbinas* (Leiden: Brill, 1992), 181.
18. Semir Zeki, *Inner Vision: An Exploration of Art and the Brain* (Oxford: Oxford University Press, 1999), 114. Richard Latto, Dennis Brain, and Brendan Kelly, "An Oblique Effect in Aesthetics: Homage to Mondrian (1872–1944)," *Perception* 29 (2000): 981–87; Richard Latto and Kirsty Russell-Duff, "An Oblique Effect in the Selection of Line Orientations by Twentieth Century Painters," *Empirical Studies of the Arts* 20 (2002): 49–60.
19. Semir Zeki and Michael Lamb, "The Neurology of Kinetic Art," *Brain* 117 (1994): 607.
20. Zeki, *Inner Vision*, 47.
21. Max Ernst, *Écritures: Avec cent vingt illustrations extraites de l'oeuvre de l'auteur* (Paris: Gallimard, 1970), 327. Translated by Samuel Fleck.
22. Ellen Dissanayake, "The Artification Hypothesis and Its Relevance to Cognitive Science, Evolutionary Aesthetics and Neuroaesthetics," *Cognitive Semiotics* 5 (2009): 156. For the artification processes from a sociological point of view, see Roberta Shapiro and Nathalie Heinich, "When Is Artification?" *Contemporary Aesthetics*, special volume 4 (2012): https://digitalcommons.risd.edu/liberalarts_contempaesthetics/volo/iss4/9.
23. Whitney Davis, "Neurovisuality," in "Evaluating Neuroaesthetics," ed. Todd Cronan, special issue, *nonsite.org*, no. 2 (2011): https://nonsite.org/neurovisuality/. See also Robert Pepperell, "Artworks as Dichotomous Objects: Implications for the Scientific Study of Aesthetic Experience," *Frontiers in Human Neuroscience* 9 (2015): https://doi.org/10.3389/fnhum.2015.00295.
24. Patrick Cavanagh, "The Artist as Neuroscientist," *Nature* 434 (March 17, 2005): 301.
25. Semir Zeki and Ludovica Marini, "Three Cortical Stages of Colour Processing in the Human Brain," *Brain* 121 (1998): 1669–85.
26. Konstantinos Moutoussis and Semir Zeki, "A Direct Demonstration of Perceptual Asynchrony in Vision," *Proceedings of the Royal Society B* 264 (1997): 393–99.
27. Vilayanur S. Ramachandran and Sandra Blakeslee, *Phantoms in the Brain* (New York: HarperCollins, 1998), 56.
28. Jean-Pierre Changeux, preface to *Raison et Plaisir* (Paris: Odile Jacob, [1989] 1994), loc. 23–30 of 2701, Kindle.
29. Zeki, *Inner Vision*, 59.
30. Susan Broadhurst, *Digital Practices: Aesthetic and Neuroesthetic Approaches to Performance and Technology* (Basingstoke, UK: Palgrave Macmillan, 2007), 57–58.
31. Gilles Deleuze, *Francis Bacon: The Logic of Sensation*, trans. Daniel W. Smith (London: Continuum [1981], 2003).
32. Semir Zeki and Tomohiro Ishizu, "The 'Visual Shock' of Francis Bacon: An Essay in Neuroesthetics," in *Bacon and the Mind: Art, Neuroscience and Psychology*, ed. Martin Harrison (London: Estate of Francis Bacon, 2019), 129, 143.
33. Marcos Nadal and Anjan Chatterjee, "Neuroaesthetics and Art's Diversity and Universality," *WIREs Cognitive Science* 10, no. 3 (2018): https://doi.org/10.1002/wcs.1487.

34. Davi Johnson Thornton, *Brain Culture: Neuroscience and Popular Media* (New Brunswick, NJ: Rutgers University Press, 2011), loc. 749 of 2829, Kindle.
35. Tomohiro Ishizu and Semir Zeki, "A Neurobiological Enquiry Into the Origins of Our Experience of the Sublime and Beautiful," *Frontiers in Human Neuroscience* 8 (2014): https://doi.org/10.3389/fnhum.2014.00891.
36. Semir Zeki, "Notes Towards a (Neurobiological) Definition of Beauty," *Gestalt Theory* 41, no. 2 (2019): 109.
37. Anjan Chatterjee, "The Neuropsychology of Visual Artistic Production," *Neuropsychologia* 42 (2004): 1573.
38. Amy Ione, "An Inquiry Into Paul Cézanne: The Role of the Artist in Studies of Perception and Consciousness," *Journal of Consciousness Studies* 7, no. 8–9 (2000): 59.
39. Beatriz Calvo-Merino, Corinne Jola, Daniel E. Glaser, and Paul Haggard, "Towards a Sensorimotor Aesthetics of Performing Art," *Consciousness and Cognition* 17 (2008): 913.
40. Zeki and Lamb, *The Neurology of Kinetic Art*, 607–8.
41. Jan B. Deregowski, "On Seeing a Picture for the First Time," *Leonardo* 9, no. 1 (1976): 19–23.
42. Erwin Panofsky, "On the Problem of Describing and Interpreting Works of the Visual Arts," [1932], *Critical Inquiry* 38, no. 3 (Spring 2012): 467–82.
43. Ione, "An Inquiry Into Paul Cézanne," 63.
44. Pierre Jacob and Marc Jeannerod, *Ways of Seeing: The Scope and Limits of Visual Cognition* (Oxford: Oxford University Press, 2003), x–xii.
45. Zeki, *Inner Vision*, 12.
46. Zeki, *Inner Vision*, 8.
47. Zeki, *Inner Vision*, 8.
48. Semir Zeki, "Neural Concept Formation and Art: Dante, Michelangelo, Wagner," in *Neurology of the Arts: Painting, Music, Literature*, ed. Frank Clifford Rose (London: Imperial College Press, 2004), 37.
49. Zeki, *Inner Vision*, 25.
50. Amy Ione, "Examining Semir Zeki's Neural Concept Formation and Art: Dante, Michelangelo, Wagner," *Journal of Consciousness Studies* 10, no. 2 (2003): 58–66.
51. Zeki, *Inner Vision*, 40.
52. Susan H. Prince, ed., *Antisthenes of Athens: Texts, Translations, and Commentary* (Ann Arbor: University of Michigan Press, 2015), 429.
53. Johann Wolfgang von Goethe, "Fortunate Encounter," in *Scientific Studies*, trans. and ed. Douglas Miller (New York: Suhrkamp, 1988), 20.
54. Ernst Cassirer, "Eidos and Eidolon: The Problem of Beauty and Art in the Dialogues of Plato," in *The Warburg Years (1919–1933): Essays on Language, Art, Myth and Technology*, trans. S. G. Lofts and A. Calcagno (New Haven, CT: Yale University Press, [1924] 2013), 231.
55. Zeki, *Inner Vision*, 40. Italics mine.
56. Margaret Livingstone, "Artisti, scienziati per istinto," in "Van Gogh e i neuroni: l'arte come esperienza cognitiva," ed. Gloria Origgi, special issue, *MicroMega*, no. 2 (2007): 194.

57. Eric R. Kandel, *Reductionism in Art and Brain Science: Bridging the Two Cultures* (New York: Columbia University Press, 2016), 118.
58. Alumit Ishai, Scott L. Fairhall, and Robert Pepperell, "Perception, Memory and Aesthetics of Indeterminate Art," *Brain Research Bulletin* 73 (2007): 319–24.
59. Margaret Livingstone, *Vision and Art: The Biology of Seeing* (New York: Harry N. Abrams, 2002), chap. 4, chap. 10.
60. William P. Seeley, "Art, Meaning, and Perception: A Question of Methods for a Cognitive Neuroscience of Art," *British Journal of Aesthetics* 53, no. 4 (2013): 452.
61. Harry Francis Mallgrave, *Architecture and Embodiment: The Implications of the New Sciences and Humanities for Design* (London: Routledge, 2013), 42. Italics mine.
62. Semir Zeki, *Balthus ou la Quête de l'Essentiel: Entretiens Avec Semir Zeki (Romans, Essais, Poésie, Documents)* (Paris: Les Belles Lettres, 1995).
63. Michael Fried, *Absorption and Theatricality: Painting and Beholder in the Age of Diderot* (Chicago: University of Chicago Press, 1980), 47.
64. Zeki, *Inner Vision*, 25.
65. Zeki, *Inner Vision*, 29.
66. Johann Wolfgang von Goethe, "Observations on the Laocoön," in *Goethe on Art*, trans. and ed. John Gage (Berkeley: University of California Press, 1980), 83.
67. Jean-Pierre Changeux, "Art and Neuroscience," *Leonardo* 27, no. 3 (1994): 199.
68. Rudolf Arnheim, *The Power of the Center: A Study of Composition in the Visual Arts* (Los Angeles: University of California Press, 1982), 52.
69. Semir Zeki, "The Neurology of Ambiguity," *Consciousness and Cognition* 13 (2004): 173–96.
70. Zeki, *Inner Vision*, 31.
71. Zeki, *Inner Vision*, 125.
72. Aage Brandt, "Form and Meaning in Art," in *The Artful Mind: Cognitive Science and the Riddle of Human Creativity*, ed. Mark Turner (Oxford: Oxford University Press, 2006), 184–85.
73. Gustav Theodor Fechner, "Vergleichende Anatomie der Engel" [1875], in *Natur der Ästhetik—Ästhetik der Natur*, ed. Olaf Breidbach (Vienna: Springer, 1997), 196–220; and Sigmund Exner, "Die Physiologie des Fliegens und Schwebens in den bildenden Künsten" [1882], in Breidbach, *Natur der Ästhetik—Ästhetik der Natur*, 225–59.
74. Olaf Breidbach, "The Beauties and the Beautiful: Some Considerations from the Perspective of Neuronal Aesthetics," in *Evolutionary Aesthetics*, ed. Eckart Voland and Karl Grammar (Berlin: Springer, 2003), 56. See also Olaf Breidbach, "Die Innenwelt der Außenwelt—Weltkonstitution im Hirngewebe? Zur Konturierung einer neuronalen Ästhetik," in *Video ergo sum: Repräsentation nach innen und außen zwischen Kunst- und Neurowissenschaften*, ed. Olaf Breidbach and Karl Clausberg, Interface 4 (Hamburg: Hans-Bredow-Institut, 1999), 34–60.
75. His view is shared by Barbara Maria Stafford, who argues that just inquiring about the body's faculty "for forming moving images or, even deeper, neural patterns—using the elementary representational tools of the sensory and motor system—entails the creation of a neuronal aesthetics." Barbara Maria Stafford, "Romantic Systematics and

the Genealogy of Thought: The Formal Roots of a Cognitive History of Images," *Configurations* 12 (2004): 316–17; Barbara Maria Stafford, "Neuronale Ästhetik: Auf dem Weg zu einer kognitiven Bildgeschichte," in *Iconic Turn: Die neue Macht der Bilder*, ed. Christa Maar and Hubert Burda (Cologne: DuMont, 2004), 103.

76. Breidbach, "The Beauties and the Beautiful," 56.
77. Catherine Malabou, *What Should We Do with Our Brain?*, trans. Sebastian Rand (New York: Fordham University Press, [2004] 2008), 13.
78. Zeki, *Inner Vision*, 68.
79. Zeki and Lamb, *The Neurology of Kinetic Art*, 607.
80. John Hyman, "Art and Neuroscience," in *Beyond Mimesis and Convention: Representation in Art and Science*, ed. Roman Frigg and Matthew C. Hunter (Heidelberg: Springer, 2010), 260.
81. Edward Wayne Massey, "Neurology and Sherlock Holmes," in *Evolutionary and Neurocognitive Approaches to Aesthetics, Creativity and the Arts*, ed. Colin Martindale, Paul Locher, and Vladimir M. Petrov (Amityville, NY: Baywood, 2007), 357–70.
82. Arnheim, *Visual Thinking*, 13.
83. Vilayanur S. Ramachandran, "Sharpening Up 'The Science of Art': An Interview with Anthony Freeman," *Journal of Consciousness Studies* 8, no. 1 (2001): 27–28.
84. Vilayanur S. Ramachandran and William Hirstein, "The Science of Art: A Neurological Theory of Aesthetic Experience," *Journal of Consciousness Studies* 6, no. 6–7 (1999): 17.
85. Ramachandran and Hirstein, "The Science of Art," 17; Ramachandran, "Sharpening Up 'The Science of Art,'" 18.
86. Vilayanur S. Ramachandran, "The Perception of Depth from Shading," *Scientific American* 234 (1976): 76, description quoted from Ramachandran and Blakeslee, *Phantoms in the Brain*, 51.
87. Ramachandran, "Sharpening Up 'The Science of Art,'" 25.
88. Ramachandran and Hirstein, "The Science of Art," 44.
89. Vilayanur S. Ramachandran, "The Artful Brain," in *A Brief Tour of Human Consciousness* (New York: Pearson, [2003] 2004), 44. For neuroscience and design, see Riccardo Falcinelli, *Guardare, pensare, progettare: neuroscienze per il design* (Viterbo, Italy: Stampa Alternativa e Graffiti, 2011).
90. Jaron Lanier, "What Information Is Given by a Veil?," *Journal of Consciousness Studies* 6, no. 6–7 (1999): 65–68.
91. Richard L. Gregory, "Object Hypotheses in Visual Perception: David Marr or Cruella de Ville?," *Journal of Consciousness Studies* 6, no. 6–7 (1999): 54–56; Richard L. Gregory, "Perceptions as Hypotheses," *Philosophical Transactions of the Royal Society B* 290, no. 1038 (1980): 181–97, in addition to "The Confounded Eye," in Richard L. Gregory and Ernst Hans Gombrich, *The Illusion in Nature and Art* (London: Duckworth, 1973), 49–96.
92. Emanuele Di Fiore and R. Di Liberti, *L'utilizzazione della stimolazione visiva subliminare nello studio dell'arte moderna* (Palermo, Italy: STASS, 1978), 26.
93. Ramachandran, "Sharpening Up 'The Science of Art,'" 18.
94. Ramachandran, "The Artful Brain," 49.

95. Ludwig Wittgenstein, "Lectures on Aesthetics," [1938], in *Lectures and Conversations: On Aesthetics, Psychology and Religious Belief*, ed. Cyril Barrett (Berkeley: University of California Press, 1966), 3.
96. Gertrude Stein, *Picasso* (London: Dover, [1938] 1984), 15.
97. Ramachandran, "The Artful Brain," 124.
98. Gilles Deleuze and Félix Guattari, *A Thousand Plateaus: Capitalism and Schizophrenia*, trans. Brian Massumi (London: Athlone, [1980] 1988), 172.
99. Chiara Cappelletto, "Weightless: Face, Faciality and Body," in *Journal of History and Philosophy of Art*, vol.1, ed. Shen Yubing (n.p.: Commercial Press, 2020), 221–42.
100. Zeki, *Inner Vision*, 54.
101. For a review of the various neuroscientific studies on inherited and acquired facial recognition and on the effects of distorted representation on the recognition of the face, see Zeki and Ishizu, "The 'Visual Shock' of Francis Bacon," 121–47; Ramachandran, "The Artful Brain," 47.
102. George Lakoff, "The Neuroscience of Form in Art," in Turner, *The Artful Mind*, 153.
103. Lakoff, "The Neuroscience of Form in Art," 164–65.
104. Vittorio Gallese and George Lakoff, "The Brain's Concepts: The Role of the Sensory-Motor System in Conceptual Knowledge," *Cognitive Neuropsychology* 22 (2005): 470–71.
105. Nicolas Poussin, *Collection de lettres de Nicolas Poussin* (Paris, [1647] 1824), 277–78. Translated by Samuel Fleck.
106. David Freedberg, "Composition and Emotion," in Turner, *The Artful Mind*, 81.
107. Ramachandran and Hirstein, "The Science of Art," 18.
108. Hyman, "Art and Neuroscience," 249.
109. Hyman, "Art and Neuroscience," 250.
110. Ramachandran and Hirstein, "The Science of Art," 16.
111. Ramachandran, "The Artful Brain," 56.
112. Richard D. Alexander, "Evolution of the Human Psyche," in *The Human Revolution: Behavioural and Biological Perspectives on the Origins of Modern Humans*, ed. Paul Mellars and Chris Stringer (Princeton, NJ: Princeton University Press, 1989), 455.
113. Alexander, "Evolution of the Human Psyche," 459.
114. Jörg Jewanski, Sean A. Day, and Jamie Ward, "A Colorful Albino: The First Documented Case of Synaesthesia, by Georg Tobias Ludwig Sachs in 1812," *Journal of the History of the Neurosciences* 18, no. 3 (2009): 293.
115. Jewanski, Day, and Ward, "A Colorful Albino," 297.
116. Maurice Merleau-Ponty, *Phenomenology of Perception*, trans. Colin Smith (London: Routledge, [1945] 2002), 266.
117. Mikel Dufrenne, *L'oeil et l'oreille* (Montreal: L'Hexagone, 1987), 65.
118. Nicholas B. Root, Karen Dobkins, Vilayanur S. Ramachandran, and Romke Rouw, "Echoes from the Past: Synaesthetic Colour Associations Reflect Childhood Gender Stereotypes," *Philosophical Transactions of the Royal Society of London B* 374 (2019): https://doi.org/10.1098/rstb.2018.0572.

119. Vilayanur S. Ramachandran and Edward M. Hubbard, "Synaesthesia: A Window Into Perception, Thought and Language," *Journal of Consciousness Studies* 8, no. 12 (2001): 3–34.
120. Ramachandran and Hubbard, "Synaesthesia," 19.
121. Emilio Gómez Milán, Oscar Iborra, María José de Cordoba, Verónica Juarez-Ramos, María Angeles Rodríguez Artacho, and José Luis Rubio, "The Kiki-Bouba Effect: A Case of Personification and Ideasthesia," *Journal of Consciousness Studies* 20, no. 2 (2013): 84–102.
122. Oren Shriki, Yaniv Sadeh, and Jamie Ward, "The Emergence of Synaesthesia in a Neuronal Network Model Via Changes in Perceptual Sensitivity and Plasticity," *PLOS Computational Biology* 12, no. 7 (2016): https://doi.org/10.1371/journal.pcbi.1004959.
123. Jerzy Konorski, *Conditioned Reflexes and Neuron Organization* (New York: Cambridge University Press, 1948), 79–80.
124. Eric R. Kandel, "Calcium and the Control of Synaptic Strength by Learning," *Nature* 293 (1981): 697–700.
125. Konorski, *Conditioned Reflexes and Neuron Organization*, 80.
126. Konorski, *Conditioned Reflexes and Neuron Organization*, 80–81.
127. Harold Burton, "Visual Cortex Activity in Early and Late Blind People," *Journal of Neuroscience* 23, no. 10 (May 2003): 4005–11.
128. Manfred Clynes and Nathan Kline, "Cyborgs and Space," *Astronautics*, September 1960.
129. Arantxa Alfaro, Ángela Bernabeu, Carlos Agulló, Jaime Parra, and Eduardo Fernánde, "Hearing Colors: An Example of Brain Plasticity," *Frontiers in Systems Neuroscience* 56, no. 9 (2015): 1–9.
130. Don Ihde and Lambros Malafouris, "Homo Faber Revisited: Postphenomenology and Material Engagement Theory," *Philosophy & Technology* 32, no. 2 (2019): 196.
131. In his talk "Cyborg Experiments," given as part of the conference "Posthuman Mimesis," hosted by Nidesh Lawtoo, Carole Guesse, and Niki Hadikoesoemo, members of the ERC project Homo Mimeticus at the Institute of Philosophy, KU Leuven, in collaboration with MDRN, the Research Group English Literature, and with the support of the European Research Council (ERC), May 20, 2021, accessed July 20, 2021, http://www.homomimeticus.eu/.
132. Andy Clark, *Natural-Born Cyborgs: Minds, Technologies, and the Future of Human Intelligence* (New York: Oxford University Press, 2004), 19.
133. Clark, *Natural-Born Cyborgs*, 28.
134. Clark, *Natural-Born Cyborgs*, 198.
135. Helmuth Plessner, *Levels of Organic Life and the Human: An Introduction to Philosophical Anthropology*, trans. Millay Hyatt (New York: Fordham University Press, [1928] 2019).
136. Etienne Bonnot de Condillac, *Traité des sensations* (Paris: A. Hatier, [1754] 1949), pt. 2, chap. 5, sec. 3. Translated by Samuel Fleck.
137. For a discussion of the Merleau-Pontian perspective in conjunction with neuroscience and the study of body maps, see Jean-Luc Petit, "La spazialità originaria del corpo

proprio Fenomenologia e neuroscienze," in *Neurofenomenologia: Le scienze della mente e la sfida dell'esperienza cosciente*, ed. Massimiliano Cappuccio (Milan: Bruno Mondadori, 2006), 163–94.

138. Quoted in Aura Satz, "'The Conviction of Its Existence': Silas Weir Mitchell, Phantom Limbs and Phantom Bodies in Neurology and Spiritualism," in *Neurology and Modernity*, ed. Laura Salisbury and Andrew Shail (Basingstoke, UK: Palgrave Macmillan, 2010), 114.
139. Jorge Luis Borges, "On Exactitude in Science" [1960], *Collected Fictions*, trans. Andrew Hurley (New York: Penguin, 1999), 325.
140. Bellamy Printz, Jaron Lanier, and Vilayanur S. Ramachandran, *Truth, Technology and the Visual/Virtual World* (Cleveland, OH: Cleveland Public Library, 2006), 31.
141. Vilayanur S. Ramachandran and Diane Rogers-Ramachandran, "Synaesthesia in Phantom Limbs Induced with Mirrors," *Proceedings of the Royal Society B*, 263, no. 1369 (1996): 377–86.
142. H. Henrik Ehrsson, Birgitta Rosén, Anita Stockselius, Christina Ragnö, Peter Köhler, and Göran Lundborg, "Upper Limb Amputees Can Be Induced to Experience a Rubber Hand as Their Own," *Brain* 131, no. 12 (December 2008): 3443–52.
143. Aviva I. Goller, Kerrie Richards, Steven Novak, and Jamie Ward, "Mirror-Touch Synaesthesia in the Phantom Limbs of Amputees," *Cortex* 49, no. 1 (2013): 243–51.
144. Vilayanur S. Ramachandran and David Brang, "Sensations Evoked in Patients with Amputation from Watching an Individual Whose Corresponding Intact Limb Is Being Touched," *Archives of Neurology* 66, no. 10 (2009): 1283.
145. Gilbert Simondon, *On the Mode of Existence of Technical Objects* , trans. **Cecile Malaspina and John Rogove** (Minneapolis: University of Minnesota Press, [1958] 2017); Karen Barad, *Meeting the Universe Halfway: Quantum Physics and the Entanglement of Matter and Meaning* (Durham, NC: Duke University Press, 2007).
146. Broadhurst, *Digital Practices*, 51.
147. Clark, *Natural-Born Cyborgs*, 115–16.
148. Pascal Giraux and Angela Sirigu, "Les mains dans la tête," *La Recherche*, no. 366 (July/August 2003): 60–63.
149. Ramachandran and Blakeslee, *Phantoms in the Brain*, 46.
150. Didier Anzieu, *The Skin Ego*, trans. Naomi Segal (London: Routledge, [1995] 2018).

## 3. NEUROARTHISTORY: ON EMOTIONS, MATTER, AND TIME

1. Gaius Plinius Secundus (Pliny the Elder), "Book XXXV: Painting, Sculpture and Architecture," in *Natural History: A Selection*, trans. John F. Healy (London: Penguin, 1992), 333–34.
2. Gottfried Boehm, "Die Wiederkehr der Bilder," in *Was ist ein Bild?*, ed. Gottfried Boehm (Munich: Fink, 1994), 11–38.
3. William P. Seeley, "Naturalizing Aesthetics: Art and the Cognitive Neuroscience of Vision," *Journal of Visual Art Practice* 5, no. 3 (2006): 210.

4. Victoria Pitts-Taylor, *The Brain's Body: Neuroscience and Corporeal Politics* (Durham, NC: Duke University Press, 2016), 51.
5. Marcos Nadal and Anjan Chatterjee, "Neuroaesthetics and Art's Diversity and Universality," *WIREs Cognitive Science* 10, no. 3 (2018): https://doi.org/10.1002/wcs.1487.
6. R. Chris Miall and John Tchalenko, "A Painter's Eye Movements: A Study of Eye and Hand Movement During Portrait Drawing," *Leonardo* 34, no. 1 (2001): 35–40.
7. Robert L. Solso, *The Psychology of Art and the Evolution of the Conscious Brain* (Cambridge, MA: MIT Press, 2003).
8. John Onians, "Neuroarthistory: First Principles," Paper presented at the Goldsmiths Neuroaesthetics Conference in May 2005.
9. Barbara Maria Stafford, "Another Kind of Global Thinking," in *Is Art History Global?*, ed. James Elkins (New York: Routledge, 2007), 184.
10. Heinrich Wölfflin, "Prolegomena to a Psychology of Architecture" [1886], in *Empathy, Form, and Space: Problems in German Aesthetics, 1873–1893*, trans. and ed. Harry Francis Mallgrave and Eleftherios Ikonomou (Santa Monica, CA: Getty Center for the History of Art and the Humanities, 1993), 184.
11. Aby Warburg, "Francesco Sassetti's Last Injunction to His Sons," in *The Renewal of Pagan Antiquity: Contributions to the Cultural History of the European Renaissance*, trans. David Britt (Los Angeles: Getty Research Institute for the History of Art and the Humanities, [1907] 1999), 249.
12. Barbara Maria Stafford, "Romantic Systematics and the Genealogy of Thought: The Formal Roots of a Cognitive History of Images," *Configurations* 12 (2004): 318.
13. Charles Darwin, *The Expression of Emotion in Man and Animals*, with introduction, afterword, and commentaries by Paul Ekman, 4th ed. (Oxford: Oxford University Press, [1872] 2009), chap. 1, 34–35.
14. Chiara Cappelletto, "The Mask: Expressing the Scattered Human Identity," in *Okologien des Ausdrucks*, ed. Franz Fehrenbach and Matthew Vollgraf (Berlin: De Gruyter, 2022).
15. Paul Ekman, afterword to *The Expression of Emotion in Man and Animals*, by Charles Darwin (Oxford: Oxford University Press, [1872] 2009), 373.
16. Jean-Jacques Courtine and Claudine Haroche, *Histoire du visage: exprimer et taire ses émotions (du XVI siècle au début du XIXe siècle)* (Paris: Payot et Rivages, [1988] 1994), 39.
17. Giovanna Colombetti, *The Feeling Body: Affective Science Meets the Enactive Mind* (Cambridge, MA: MIT Press, 2014).
18. James Elkins, *Pictures and Tears* (New York: Routledge, [2001] 2005), 28.
19. Elkins, *Pictures and Tears*, vii.
20. Gotthold Ephraim Lessing, *Laocoön: An Essay on the Limits of Painting and Poetry*, trans. Edward Allen McCormick (Baltimore: Johns Hopkins University Press, [1766] 1984), sec. 3.
21. Elkins, *Pictures and Tears*, 68.
22. Elkins, *Pictures and Tears*, 71.
23. Semir Zeki, *Inner Vision: An Exploration of Art and the Brain* (Oxford: Oxford University Press, 1999), 51.

24. David Freedberg, "Empathy, Motion and Emotion," in *Wie sich Gefühle Ausdruck verschaffen: Emotionen in Nahsicht*, ed. Klaus Herding and Antje Krause Wahl (Berlin: Driesen, 2007), 31.
25. David Freedberg, "The Problem of Classicism: Ideology and Power," *Art Journal* 47, no. 1 (Spring 1988): 8; Freedberg, "Empathy, Motion and Emotion," 21.
26. Stafford, "Another Kind of Global Thinking," 186.
27. Antonio R. Damasio, *Descartes' Error: Emotion, Reason, and the Human Brain*, reprint ed. (New York: Penguin, [1994] 2005), 184.
28. William James, *Principles of Psychology* (New York: Dover, [1890] 1950), 2:449.
29. Edmund Burke, *A Philosophical Enquiry Into the Origin of Our Ideas of the Sublime and Beautiful*, ed. Paul Guyer (Oxford: Oxford University Press, [1757] 1990), pt. 1, chap. 8, 38.
30. David Freedberg, *The Power of Images: Studies in the History and Theory of Response* (Chicago: University of Chicago Press, 1989), xx.
31. Freedberg, *The Power of Images*, 430.
32. Freedberg, *The Power of Images*, 432.
33. Freedberg, "Empathy, Motion and Emotion," 18.
34. Freedberg, *The Power of Images*, 437.
35. Freedberg, *The Power of Images*, xxv.
36. Moritz Geiger, "Vom Dilettantismus im künstlerischen Erleben," in *Zugänge zur Ästhetik* (Leipzig: Der Neue Geist, 1928), 7.
37. Bernard Berenson, *Seeing and Knowing* (n.p.: Forgotten Books, [1953] 2017), 55.
38. Berenson, *Seeing and Knowing*, 62.
39. Ernst Hans Gombrich, "Concerning 'The Science of Art': Commentary on Ramachandran and Hirstein," *Journal of Consciousness Studies* 7, no. 8–9 (2000): 17; and see the reply by Vilayanur S. Ramachandran, "Response to Gombrich," *Journal of Consciousness Studies* 7, no. 8–9 (2000): 17–20.
40. Freedberg, "Empathy, Motion and Emotion," 42.
41. Jean-Pierre Changeux, "Art and Neuroscience," *Leonardo* 27, no. 3 (1994): 193.
42. Changeux, "Art and Neuroscience," 196.
43. Changeux, "Art and Neuroscience," 189.
44. See the catalogue *Les passions de l'âme: Peintures des XVIIe et XVIIIe siècles de la collection Changeux*, published in conjunction with exhibitions at the Musée Bossuet de Meaux, the Musée des Augustins de Toulouse, and the Musée des Beaux-Arts de Caen in 2006, with pieces by Jean-Pierre Changeux, Nicole Rouillé, and Jean-Claude Boyer (Paris: Odile Jacob, 2006). See also Jean-Pierre Changeux, *Raison et Plaisir* (Paris: Odile Jacob, [1989] 1994), loc. 933–1218 of 2701, Kindle.
45. Irina Vladimirovna Linnik, "Un tableau de Jacques de Bellange nouvellement découvert," *Revue de l'Art* 20 (1973): 68–69.
46. Changeux, "Art and Neuroscience," 196.
47. Changeux, "Art and Neuroscience," 190.
48. Changeux, *Raison et Plaisir*, loc. 1054 of 2701, Kindle.
49. Changeux, "Art and Neuroscience," 196.
50. Changeux, "Art and Neuroscience," 194.

51. Changeux, "Art and Neuroscience," 193.
52. Nelson Goodman, *The Languages of Art* (Indianapolis: Hackett, 1968), 247–48.
53. John Onians, *Neuroarthistory: From Aristotle and Pliny to Baxandall and Zeki* (New Haven, CT: Yale University Press, 2007), 17.
54. John Onians, "The Biological Basis of Renaissance Aesthetics," in *Concepts of Beauty in Renaissance Art*, ed. Francis Ames-Lewis and Mary Rogers (Aldershot, UK: Ashgate, 1998), 12.
55. John Onians, "Francis Bacon: A Neuroarthistory," in *Bacon and the Mind: Art, Neuroscience and Psychology* (London: Estate of Francis Bacon, 2019), 100.
56. Whitney Davis, "Neurovisuality," in "Evaluating Neuroaesthetics," ed. Todd Cronan, special issue, *nonsite.org*, no. 2 (2011): https://nonsite.org/neurovisuality/.
57. Jay Appleton, *The Experience of Landscape* (New York: Wiley, 1996), vii.
58. Onians, "The Biological Basis of Renaissance Aesthetics," 15.
59. Matthew Rampley, "Fish, Volcanoes and the Art of Brains: Review of John Onians European Art: A Neuroarthistory," *Journal of Art Historiography* 15 (2016), accessed March 3, 2021, https://arthistoriography.files.wordpress.com/2016/10/rampley-review-2.pdf.
60. John Onians, "Greek Temple and Greek Brain," in *Body and Building: Essays on the Changing Relation of Body and Architecture*, ed. George Dodds and Robert Tavernor (Cambridge, MA: MIT Press, 2002), 43–63.
61. Davis, "Neurovisuality." Italics mine.
62. W. J. T. Mitchell, "Showing Seeing: A Critique of Visual Culture," in *Art History, Aesthetics, Visual Studies*, ed. Michael Ann Holly and Keith Moxey (Williamstown, MA: Clark Art Institute, 2002), 236. Quoted in John Onians, *European Art: A Neuroarthistory* (New Haven, CT: Yale University Press, 2016), 4.
63. Onians, *European Art*, 3. Italics mine.
64. Clement Greenberg, "Avant-Garde and Kitsch," [1939], in *Art and Culture: Critical Essays* (Boston: Beacon Press, 1961), 12.
65. Vilayanur S. Ramachandran, "The Artful Brain," in *A Brief Tour of Human Consciousness* (New York: Pearson, [2003] 2004), 129–30.
66. Colin Martindale, "Peak Shift, Prototypicality and Aesthetic Preference," *Journal of Consciousness Studies* 6, no. 6–7 (1999): 52–54.
67. Greenberg, "Avant-Garde and Kitsch," 14.
68. Lamberto Maffei and Adriana Fiorentini, *Arte e cervello* (Bologna: Zanichelli, 1995), 177.
69. John Onians, "A Brief Natural History of Art," in *Compression vs. Expression: Containing and Explaining the World's Art*, ed. John Onians (Williamstown, MA: Clark Art Institute, 2000), 235–49.
70. Onians, "A Brief Natural History of Art," 239–40.
71. James Elkins, "Art History as a Global Discipline," in Elkins, *Is Art History Global?*, 19–21; Matthew Rampley, "From Big Art Challenge to a Spiritual Vision: What 'Global Art History' Might Really Mean," in Elkins, *Is Art History Global?*, 188–202.
72. John Onians, "A Natural Anthropology of Art," *International Journal of Anthropology* 18, no. 4 (2003): 261.
73. Steven Pinker, *How the Mind Works* (New York: Norton, 1997), 534.

74. Ernst Hans Gombrich, *Art and Illusion: A Study in the Psychology of Pictorial Representation* (Princeton, NJ: Princeton University Press, 1960), 298.
75. Onians, *Neuroarthistory: From Aristotle and Pliny to Baxandall and Zeki*, 10.
76. Onians, *Neuroarthistory: From Aristotle and Pliny to Baxandall and Zeki*, 29. On the place of Aristotle in neuroscientific research, see Tullio Manzoni, *Aristotele e il cervello: le teorie del più grande biologo dell'antichità nella storia del pensiero scientifico* (Rome: Carocci, 2007).
77. Onians, *Neuroarthistory: From Aristotle and Pliny to Baxandall and Zeki*, 29.
78. Onians, *Neuroarthistory: From Aristotle and Pliny to Baxandall and Zeki*, 78.
79. André Malraux, "The Museum Without Walls," in *The Voices of Silence*, trans. Stuart Gilbert (Princeton, NJ: Princeton University Press, 1978), 85.
80. Quoted in Vilayanur S. Ramachandran and Sandra Blakeslee, *Phantoms in the Brain* (New York: HarperCollins, 1998), 123–24.
81. Onians, *A Brief Natural History of Art*, 248.
82. Hal Foster, Rosalind Krauss, Yve-Alain Bois, and Benjamin H. D. Buchloh, *Art Since 1900: Modernism, Antimodernism, Postmodernism, Vol. 2: 1945 to the Present*, reprinted with corrections (New York: Thames and Hudson, 2007), 67.
83. Ramachandran, "The Artful Brain," 130.
84. Davis, "Neurovisuality."
85. Ernst Hans Gombrich, *The Sense of Order: A Study in the Psychology of Decorative Art* (London: Phaidon, 1979), xii.
86. Ernst Hans Gombrich and Didier Eribon, *Looking for Answers: Conversations on Art and Science* (New York: Harry N. Abrams, [1991] 1993), 174.
87. Gombrich and Eribon, *Looking for Answers*, 133.
88. Gombrich, *Art and Illusion*, 102.
89. Onians, *Neuroarthistory: From Aristotle and Pliny to Baxandall and Zeki*, 65.
90. Karl Clausberg, *Neuronale Kunstgeschichte: Selbstdarstellung als Gestaltungsprinzip* (Vienna: Springer, 1999).
91. Olaf Breidbach, "The Beauties and the Beautiful: Some Considerations from the Perspective of Neuronal Aesthetics," in *Evolutionary Aesthetics*, ed. Eckart Voland and Karl Grammar (Berlin: Springer, 2003), 39–68.
92. Berenson, *Seeing and Knowing*, 27–28.
93. Barbara Maria Stafford, *Echo Objects: The Cognitive Work of Images* (Chicago: University of Chicago Press, 2007), 25.
94. Stafford, "Romantic Systematics," 344.
95. Stafford, *Echo Objects*, 29.
96. Stafford, "Romantic Systematics," 335.
97. Stafford, *Echo Objects*, 35–36.
98. Stafford, "Romantic Systematics," 317.
99. Rudolf Reicke, ed., *Lose Blätter aus Kants Nachlass* (Königsberg: Beyer, 1895), 119.
100. Giambattista Vico, *The New Science of Giambattista Vico: Unabridged Translation of the Third Edition, with the Addition of "Practic of the New Science,"* trans. Thomas Goddard Bergin and Max Harold Fisch (Ithaca, NY: Cornell University Press, [1774] 2015), bk. 1, sec. 3–4.
101. Nadal and Chatterjee, "Neuroaesthetics and Art's Diversity and Universality."

102. Stafford, *Echo Objects*, 3.
103. Stafford, "Romantic Systematics," 319.
104. Humberto R. Maturana and Francisco J. Varela, *Autopoiesis and Cognition: The Realization of the Living* (Dordrecht, Netherlands: D. Reidel, 1980), 79, 135; cf. Stafford, *Echo Objects*, 175–204.
105. Antonio R. Damasio, "L'omeostasi artistica," in "Van Gogh e i neuroni: L'arte come esperienza cognitiva," ed. Gloria Origgi, special issue, *MicroMega* 2 (2007): 187–88.
106. Johann Gottfried Herder, *Sculpture: Some Observations on Shape and Form from Pygmalion's Creative Dream*, trans. Jason Gaiger (Chicago: University of Chicago Press, [1778] 2002), chap. 1, sec. 2, 37–38.
107. Stafford, *Echo Objects*, 10.
108. Gombrich, *Art and Illusion*, 86.
109. Stafford, *Echo Objects*, 77.

## 4. NEUROARTCRITICISM: FROM THE ARTIST'S LESIONS TO THE ARTWORK AND VICE VERSA

1. Hanna Damasio, Thomas Grabowski, Randall Frank, Albert M. Galaburda, and Antonio R. Damasio, "The Return of Phineas Gage: Clues About the Brain from the Skull of a Famous Patient," *Science* 264, no. 5162 (1994): 1102–5.
2. Tony James, *Dream, Creativity, and Madness in Nineteenth-Century France* (Oxford: Clarendon, 1995).
3. Jon O. Lauring, "The Theoretical and Methodological Backdrop of Neuroaesthetics," in *An Introduction to Neuroaesthetics: The Neuroscientific Approach to Aesthetic Experience, Artistic Creativity, and Arts Appreciation*, ed. Jon O. Lauring (Copenhagen: Museum Tusculanum, 2014), 80–81.
4. Arthur P. Shimamura, "Muybridge in Motion: Travels in Art, Psychology and Neurology," *History of Photography*, no. 26 (2002): 341–50.
5. Giorgio Vasari, "A Description of the Works of Titian of Cadore, Painter," in *The Lives of the Artists*, trans. Julia Conaway Bondanella and Peter Bondanella (New York: Oxford University Press, [1550] 1991), 507–8.
6. Vasari, "A Description of the Works of Titian of Cadore, Painter," 502–3.
7. Antonio Palomino de Castro y Velasco, "Don Diego Velázquez de Silva," in *El Museo pictórico y escala óptica* (Madrid: Aguilar, [1796] 1988), vol. 3, sec. 3, 225.
8. Robert Zwijnenberg, "Brains, Art and the Humanities," in *Neurocultures: Glimpses Into an Expanding Universe*, ed. Francisco Ortega and Fernando Vidal (Oxford: Peter Lang, 2011), 303.
9. Jonah Lehrer, "Unlocking the Mysteries of the Artistic Mind," *Psychology Today*, July 1, 2010.
10. Giandomenico Iannetti, email message to the author, February 3, 2020.
11. Dahlia W. Zaidel, *Neuropsychology of Art: Neurological, Cognitive and Evolutionary Perspectives* (Hove, UK: Psychology Press, 2005).

12. Anna Mazzucchi, Giovanni Pesci, and Dario Trento, *Cervello e pittura* (Rome: Fratelli Palombi Editori, 2004), 312.
13. Anna Berti and Alessia Folegatti, "Negligenza spaziale: Immagini, distorsioni e modulazioni delle rappresentazioni spaziali nella produzione grafica e artistica di pazienti cerebrolesi," in *Immagini della mente: Neuroscienze, arte, filosofia*, ed. Giovanni Lucignani and Andrea Pinotti (Milan: Cortina, 2007), 207–29.
14. Olaf Blanke, "Visuospatial Neglect in Lovis Corinth's Self-Portraits," in *The Neurobiology of Painting*, ed. Frank Clifford Rose (San Diego: Elsevier, 2006), 212.
15. Mazzucchi, Pesci, and Trento, *Cervello e pittura*, 304.
16. Anna Cantagallo and Sergio Della Sala, "Preserved Insight in an Artist with Extrapersonal Spatial Neglect," *Cortex* 34, no. 2 (1998): 180.
17. Eric R. Kandel, James H. Schwartz, and Thomas M. Jessell, *Principles of Neural Science*, 5th ed. (New York: McGraw-Hill, 2013), 38; the authors cite the definition from Jerzy Konorski, *Conditioned Reflexes and Neuron Organization* (New York: Cambridge University Press, 1948).
18. Betty Edwards, *Drawing on the Right Side of the Brain*, 4th ed. (New York: Tarcher/Penguin, [1979] 2012), loc. 145 of 5140, loc. 99 of 5140, Kindle.
19. Robert Hughes, *Goya* (New York: Knopf, 2003), 138. See also Frank Clifford Rose, "The Deafness of Goya (1746–1828)," in Rose, *The Neurobiology of Painting*, 301–15.
20. Henri Gastaut, "La maladie de Vincent van Gogh envisagée à la lumière des conceptions nouvelles sur l'épilepsie psychomotrice," *Annales Médico-Psychologiques* 114 (1956): 196–238.
21. Tiina Rekand, "Absinthe: The Nervous System and Painting," in Rose, *The Neurobiology of Painting*, 271–78.
22. Antonin Artaud, "Van Gogh, the Man Suicided by Society," [1947], in *Antonin Artaud: Anthology*, ed. Jack Hirschman (San Francisco: City Lights, 1965), 139.
23. Fyodor M. Dostoevsky, *The Idiot*, trans. David McDuff (London: Penguin, [1869] 2004), 264. For a history of the author's epilepsy, see Halfdan Kierulf, "The Aetiology of Dostoevsky's Epilepsy," in *Neurology of the Arts: Painting, Music, Literature*, ed. Frank Clifford Rose (London: Imperial College Press, 2004), 351–56.
24. Fabio Cirignotta, C. V. Todesco, and Elio Lugaresi, "Temporal Lobe Epilepsy with Ecstatic Seizures (So-Called Dostoevsky Epilepsy)," *Epilepsia* 21 (1980): 705–10.
25. William G. Lennox, *Epilepsy and Related Disorders*, with Margaret A. Lennox (London: Churchill, 1960), 2 vols., of which see vol. 2, chap. 21, "Epileptics of Worth and Fame."
26. Oliver Sacks, "The Landscape of His Dreams," in *An Anthropologist on Mars: Seven Paradoxical Tales* (New York: First Vintage, [1995] 1996).
27. Anjan Chatterjee, "The Neuropsychology of Visual Art: Conferring Capacity," in Rose, *The Neurobiology of Painting*, 39–40.
28. Cecilia Serrano, Richard Allegri, Mario Martelli, Fernando Teragano, and P. Rinalli, "Visual Art, Creativity and Dementia," *Vertex* 16 (2005): 418–29; Valeria Drago, Paul S. Foster, Daniel Trifiletti, David B. FitzGerald, Gregory P. Crucian, and Kenneth M. Heilman, "'What's Inside the Art?' The Influence of Frontotemporal Dementia in Art Production," *Neurology* 10 (2006): 1285–87.

29. Carlos Hugo Espinel, "De Kooning's Late Colours and Forms: Dementia, Creativity and the Healing Power of Art," *Lancet* 347 (1996): 1096–98; Gary Garrels, "Three Toads in the Garden: Line, Color and Form," in *Willem de Kooning: The Late Paintings, the 1980s* (Minneapolis: San Francisco Museum of Modern Art and Walker Art Center, 1995), exhibition catalogue.
30. J. Allan Hobson and Hellmut Wohl, *Dagli angeli ai neuroni: l'arte e la nuova scienza dei sogni* (Fidenza, Italy: Mattioli 1885, [2005] 2007), 48.
31. Julien Bogousslavsky, "Creativity in Painting and Style in Brain-Damaged Artists," in Rose, *The Neurobiology of Painting*, 142–43; Bruce L. Miller and Craig E. Hou, "Portraits of Artists: Emergence of Visual Creativity in Dementia," *Archives of Neurology* 61 (2004): 842–44.
32. Sebastian J. Crutch and Martin N. Rossor, "Artistic Changes in Alzheimer's Disease," in Rose, *The Neurobiology of Painting*, 160.
33. Anjan Chatterjee and Oshin Vartanian, "Neuroaesthetics," *Trends in Cognitive Sciences* 18, no. 7 (2014): 373.
34. James Croft, "The Challenges of Interdisciplinary Epistemology in Neuroaesthetics," *Mind, Brain and Education* 5, no. 1 (2011): 9.
35. Stefano F. Cappa, "Arte e immagine corporea: considerazioni neurologiche," in Lucignani and Pinotti, *Immagini della mente*, 121–34.
36. Angela Wagner, Matthias Ruf, Dieter F. Braus, and Martin H. Schmidt, "Neuronal Activity Changes and Body Image Distortions in Anorexia Nervosa," *NeuroReport* 14 (2003): 2193–97.
37. David Katz, *War Greco astigmatisch? Eine psychologische Studie zur Kunstwissenschaft* (Berlin: De Gruyter, [1914] 2021).
38. Espen Dietrichs, "Peg Leg Frieda," in Rose, *The Neurobiology of Painting*, 293–99.
39. Robert L. Solso, *The Psychology of Art and the Evolution of the Conscious Brain* (Cambridge, MA: MIT Press, 2003), 13.
40. Solso, *The Psychology of Art*, 57.
41. Solso, *The Psychology of Art*, 253.
42. Bogousslavsky, *Creativity in Painting and Style in Brain-Damaged Artists*, 136.
43. Gaius Plinius Secundus (Pliny the Elder), "Book XXXV: Painting, Sculpture and Architecture," in *Natural History: A Selection*, trans. John F. Healy (London: Penguin, 1992), 332.
44. Wassily Kandinsky, *Point and Line to Plane*, trans Howard Dearstyne (New York: Dover, [1926] 1979).
45. Eric R. Kandel, James H. Schwartz, and Thomas M. Jessell, *Principles of Neural Science*, 3rd ed. (Norwalk, CT: Appleton and Lange, 1991), 367.
46. Kay Redfield Jamison, *Touched with Fire: Manic-Depressive Illness and the Artistic Temperament* (New York: Simon & Schuster, 1993).
47. Sacks, "The Landscape of His Dreams," 85.
48. Lorna Selfe, *Nadia: A Case of Extraordinary Drawing Ability in an Autistic Child* (Cambridge, MA: Academic Press, 1977).
49. Lamberto Maffei and Adriana Fiorentini, *Arte e cervello* (Bologna: Zanichelli, 1995), 156–58.

50. Amir Amedi, Lofti B. Merabet, Noa Tal, and Alvaro Pascual-Leone, "Pictorial Art Beyond Sight: Revealing the Mind of a Blind Painter," in *Art and the Senses*, ed. Francesca Bacci and David Melcher (New York: Oxford University Press, 2011), 468.
51. Maffei and Fiorentini, *Arte e cervello*, 156–57.
52. Rudolf Arnheim, "The Art of Psychotics," *Art Psychotherapy* 4 (1977): 114.
53. Hans Prinzhorn, *The Artistry of the Mentally Ill: A Contribution to the Psychology and Psychopathology of Configuration*, trans. Eric von Brockdorff (New York: Springer, [1922] 1972), xviii.
54. Prinzhorn, *The Artistry of the Mentally Ill*, 12.
55. Prinzhorn, *The Artistry of the Mentally Ill*, 16.
56. Howard Gardner and Ellen Winner, "Children's Conceptions (and Misconceptions) of the Arts," in *Art, Mind, and Brain: A Cognitive Approach to Creativity* (New York: Basic Books, 1982), 104–9.
57. Felix Klee, ed., *Paul Klee: His Life and Work in Documents, Selected from Posthumous Writings and Unpublished Letters*, trans. Richard and Clara Winston (New York: Braziller, [1960] 1962), 183.
58. Paul Klee, *The Diaries of Paul Klee 1898–1918*, ed. Felix Klee (Berkeley: University of California Press, [1957] 1968), sec. 905, 266.
59. Hal Foster, Rosalind Krauss, Yve-Alain Bois, and Benjamin H. D. Buchloh, *Art Since 1900: Modernism, Antimodernism, Postmodernism, Vol. 1: 1900 to 1944* (New York: Thames and Hudson, 2004), 55–56.
60. Jean-Michel Palmier, François Aubral, Anthony Rowley, Pierre Vallaud, and Jean-Nöel von der Weid, *L'art dégénéré: Une exposition sous le IIIe Reich* (Paris: Jacques Bertoin, [1992] 2002), 99.
61. Tobin Siebers, *Disability Aesthetics* (Ann Arbor: University of Michigan Press, 2010), 35.
62. Harry Francis Mallgrave, *Architecture and Embodiment: The Implications of the New Sciences and Humanities for Design* (London: Routledge, 2013), 28.
63. Thomas Sebeok, "Prefigurements of Art," in *Readings in Zoosemiotics*, ed. Timo Maran, Dario Martinelli, and Aleksei Turovski (Berlin: De Gruyter, 2011), 220.
64. Desmond Morris, *The Biology of Art: A Study of the Picture-Making Behavior of the Great Apes and Its Relationship to Human Art* (London: Methuen, 1969), 150.
65. Morris, *The Biology of Art*, 158.
66. Lamberto Maffei, "I diversi sentieri della memoria e l'arte visiva," in Lucignani and Pinotti, *Immagini della mente*, 90. Italics mine. See Catherine M. Soussloff, "Jackson Pollock's Post-Ritual Performance: Memories Arrested in Space," *Drama Review* 48, no. 1 (T181) (Spring 2004): 60–78. See also Robert Goodnough's 1951 account, "Pollock Paints a Picture," *Art News*, November 26, 2012, accessed February 11, 2020, https://www.artnews.com/art-news/news/pollock-paints-a-picture-2127.
67. William P. Seeley, "Art, Meaning, and Aesthetics: The Case for a Cognitive Neuroscience of Art," in *Art, Aesthetics, and the Brain*, ed. Marcos Nadal, Joseph P. Huston, Luigi Agnati, Francisco Mora, and Camilo José Cela-Conde (New York: Oxford University Press, 2015), 26.

68. Anjan Chatterjee, "The Neuropsychology of Visual Artistic Production," *Neuropsychologia* 42 (2004): 1568.
69. Ernst Hans Gombrich, "Psychoanalysis and the History of Art," in *Meditations on a Hobby Horse and Other Essays* (London: Phaidon, 1963), 31.
70. Hobson and Wohl, *Dagli angeli ai neuroni*, 121.
71. Frank Clifford Rose, "The Neurology of Art: An Overview," in *Neurology of the Arts: Painting, Music, Literature*, 60. See also Manooj Ramachandran and Jeffrey K. Aronson, "The Diagnosis of Art: Arthrogryposis and Ribera's *The Clubfoot*," *Journal of the Royal Society of Medicine* 99, no. 6 (2006): 321–22.
72. Alan Eglin Heathcote Emery, "Neurogenetics in Art," in Rose, *The Neurobiology of Painting*, 239.
73. Friedrich Nietzsche, *The Dawn of Day*, trans. Johanna Volz (London: T. Fisher Unwin, [1879–81] 1903), sec. 8.
74. Philip Smith, "Fainting in Classical Art," in Rose, *The Neurobiology of Painting*, 79–88.
75. Derek Robinson and Klaus Podoll, *Migraine Art: The Migraine Experience from Within* (Berkeley, CA: North Atlantic, 2009).
76. Klaus Podoll, "Migraine Art in the Internet: A Study of 450 Contemporary Artists," in Rose, *The Neurobiology of Painting*, 89–107.
77. Ubaldo Nicola and Klaus Podoll, *L'aura di Giorgio de Chirico: Arte emicranica e pittura metafisica* (Milan: Mimesis, 2003), 13.
78. Giorgio De Chirico, "Sull'arte metafisica," in *Il meccanismo del pensiero: Critica, polemica, autobiografia. 1911–1943*, ed. Maurizio Fagiolo (Turin: Einaudi, 1985), 85–86.
79. Giovanni Morelli, *The Borghese and Doria-Pamfili Galleries in Rome*, trans. Constance Joyce Ffoulkes (London: J. Murray, [1897] 1988), 37. For Morelli's place in a "psychobiological" tradition that includes Darwin and Wöllflin, see Lauren Golden, "Science, Darwin and Art History," in *Raising the Eyebrow: John Onians and World Art Studies: An Album Amicorum in His Honour*, ed. Lauren Golden (Oxford: Archaeopress, 2001), 79–90.
80. Sigmund Freud, "The Moses of Michelangelo," in *The Standard Edition of the Complete Works of Sigmund Freud*, trans. and ed. James Strachey (London: Vintage, [1914] 2001), 13:230.
81. Rose-Carol Washton Long, ed., "Franz Marc, letter to Reinhard Piper, April 20, 1910, and to Wassily Kandinsky, October 5, 1912," *German Expressionism: Documents from the End of the Wilhelmine Empire to the Rise of National Socialism*, with the assistance of Ida Katherine Rigby and contributions by Stephanie Barron, Rosemarie Haag Bletter, and Peter Chametzky, trans. and ed. Nancy Roth (Berkeley: University of California Press, 1995), 49.
82. Charles Robert Leslie, *Memoirs of the Life of John Constable Composed Chiefly of His Letters*, ed. Jonathan Mayne (London: Phaidon, 1951), 323.
83. Hermann Rorschach, *Psychodiagnostics: A Diagnostic Test Based on Perception*, trans. Paul Lemkau, 2nd ed. [1932] (New York: Grune & Stratton, [1921] 1942).
84. Leonardo Da Vinci, *A Treatise on Painting* [1540], trans. John Francis Rigaud (London: George Bell and Sons, 1892), 149–50.

85. Ernst Hans Gombrich, *Art and Illusion: A Study in the Psychology of Pictorial Representation* (Princeton, NJ: Princeton University Press, 1960), 151.
86. Nicola and Podoll, *L'aura di Giorgio de Chirico*, 126.
87. Waldemar Conrad, "Der ästhetische Gegenstand: Eine phänomenologische Studie," *Zeitschrift für Ästhetik und Allgemeine Kunstwissenschaft* 3 (1908): 71–72.
88. Chatterjee, *The Neuropsychology of Visual Art*, 47.
89. Anjan Chatterjee, "Neuroaesthetics: A Coming of Age Story," *Journal of Cognitive Neuroscience* 23, no. 1 (2011): 57.
90. Catherine Malabou, *Ontology of the Accident: An Essay on Destructive Plasticity*, trans. C. Shread (Cambridge: Polity, [2009] 2012), 30.

## 5. THE BRAIN'S ICONOCLASH

1. Tomoyasu Horikawa and Yukiyasu Kamitani, "Generic Decoding of Seen and Imagined Objects Using Hierarchical Visual Features," *Nature Communications* 8 (2017): https://doi.org/10.1038/ncomms15037; "Kamitani Lab," Department of Information Science and Technology, Kyoto University, accessed June 10, 2021, https://kamitani-lab.ist.i.kyoto-u.ac.jp/page/10.
2. Amit Prasad, "Making Images/Making Bodies: Visibilizing and Disciplining Through Magnetic Resonance Imaging (MRI)," *Science, Technology, & Human Values* 30 (2005): 304.
3. Ksenia Fedorova, "Neurointerfaces, Mental Imagery and Sensory Translation in Art and Science in the Digital Age," in *Invisibility in Visible and Material Culture*, ed. Asbjørn Grønstad and Øyvind Vågnes (n.p.: Palgrave Macmillan, 2019), 94. Italics mine.
4. Patricia S. Churchland, *Touching a Nerve: The Self as Brain* (New York: Norton, 2013), 11.
5. "SPECT Gallery: A Look Inside the Mind with Brain SPECT Imaging," Amen Clinics, accessed June 10, 2021, https://www.amenclinics.com/gallery/.
6. Anne Beaulieu, "Images Are Not the (Only) Truth: Brain Mapping, Visual Knowledge, and Iconoclasm," *Science, Technology, & Human Values* 27, no. 1 (2002): 62.
7. Beaulieu, "Images Are Not the (Only) Truth," 74.
8. Beaulieu, "Images Are Not the (Only) Truth," 55.
9. Jonathan Crary, *Techniques of the Observer: On Vision and Modernity in the Nineteenth Century* (Cambridge, MA: MIT Press, 1990), 1.
10. Antonio R. Damasio, "Body Images," in *The Introduction to Images: A Reader*, ed. Sunil Manghani, Arthur Piper, and Jon Simons (London: Sage, 2006), 199–202.
11. Bruno Latour, "What Is Iconoclash? Or Is There a World Beyond the Image Wars?," in *Beyond the Image-Wars in Science, Religion and Art*, ed. Peter Weibel and Bruno Latour (Cambridge, MA: MIT Press, 2002), 19.
12. Richard Wingate and Marius Kwint, "Imagining the Brain Cell: The Neuron in Visual Culture," *Nature Reviews Neuroscience* 7, no. 9 (September 2006): 751.
13. "Brain Art Competition 2020," Neuro Bureau, accessed June 10, 2021, https://www.neurobureau.org/galleries/brain-art-competition-2020/.

14. Niels Bohr, "The Atomic Theory and the Fundamental Principles Underlying the Description of Nature," in *Atomic Theory and the Fundamental Principles Underlying the Description of Nature* (Cambridge: Cambridge University Press, [1929] 1934), 102–19.
15. Karen Barad, *Meeting the Universe Halfway: Quantum Physics and the Entanglement of Matter and Meaning* (Durham, NC: Duke University Press, 2007), 142.
16. Barad, *Meeting the Universe Halfway*, 144.
17. Barad, *Meeting the Universe Halfway*, 148.
18. Felice Frankel, *L'incanto della scienza*, ed. Maria Perosino (Turin Codice, 2004), 9.
19. Hans Christian Adam, ed., *Karl Blossfeldt 1865–1932* (Cologne: Taschen, 1999); Ernst Haeckel, *Art Forms in Nature* (Munich: Prestel, [1904] 1998).
20. Roald Hoffmann, "Making Sense of the Image in the Nanoworld," accessed June 10, 2021, http://www.roaldhoffmann.com/other.
21. Plautus, *Epid.* 382–89.
22. Diogenes Laertius, *Lives* 2.33.
23. Sabine Melchior-Bonnet, *The Mirror: A History*, trans. Katharine H. Jewett (New York: Routledge, [1995] 2009).
24. Silvia Casini, "Magnetic Resonance Imaging (MRI) as Mirror and Portrait: MRI Configurations Between Science and the Arts," *Configurations* 19, no. 1 (2011): 89.
25. Andrew P. Wickens, *A History of the Brain: From Stone Age Surgery to Modern Neuroscience* (New York: Psychology Press, 2015), 69.
26. Wickens, *A History of the Brain*, 49.
27. Laura Favaretto, "Thinking Head," accessed July 19, 2021, http://www.thinking-head.net/.
28. Wickens, *A History of the Brain*, 138.
29. Wickens, *A History of the Brain*, 171.
30. Pablo García-López, Virginia García-Marín, and Miguel Freire, "The Histological Slides and Drawings of Cajal," *Frontiers in Neuroanatomy* 4, no. 9 (2010): https://doi.org/10.3389/neuro.05.009.2010.
31. Étienne-Jules Marey, *La méthode graphique dans les sciences expérimentales et principalement en physiologie et en médecine* (Paris: G. Masson, 1878).
32. Stefano Sandrone, Marco Bacigaluppi, Marco R. Galloni, Stefano F. Cappa, Andrea Moro, Marco Catani, Massimo Filippi, et al., "Weighing Brain Activity with the Balance: Angelo Mosso's Original Manuscripts Come to Light," *Brain* 137, no. 2 (February 2014): 621–33.
33. Søren Kierkegaard, *Papers and Journals: A Selection*, trans. Alastair Hannay (New York: Penguin, [1846] 1996), 369.
34. Jameson Kismet Bell, "The Performativity of a Historical Brain Event: Revisiting 1517 Strassburg," in *The Neuroscientific Turn: Transdisciplinarity in the Age of the Brain*, ed. Melissa M. Littlefield and Jenell Johnson (Ann Arbor: University of Michigan Press, 2012), 63.
35. "The Whole Brain Atlas," Harvard Medical School, accessed June 10, 2021, https://www.med.harvard.edu/aanlib.
36. Prasad, "Making Images/Making Bodies," 293.
37. Jean Baudrillard, *America*, trans. Chris Turner (London: Verso, [1986] 1988), 36.

38. Susan M. Fitzpatrick, "Functional Brain Imaging: Neuro-Turn or Wrong Turn?," in Littlefield and Johnson, *The Neuroscientific Turn*, 186.
39. Yan Bao and Ernst Pöppel, "Anthropological Universal and Cultural Specifics: Conceptual and Methodological Challenges in Cultural Neuroscience," *Neuroscience & Biobehavioral Reviews* 36, no. 9 (2012): 2144.
40. "Brain & Behavior. Psychophysiology. Clinical Translational," Center for the Study of Emotion and Attention, University of Florida, accessed June 10, 2021, https://csea.phhp.ufl.edu/.
41. Prasad, "Making Images/Making Bodies," 292.
42. Prasad, "Making Images/Making Bodies," 298–99.
43. Prasad, "Making Images/Making Bodies," 299.
44. Beaulieu, "Images Are Not the (Only) Truth," 64.
45. As shown in Marta Calbi, Monica Angelini, Vittorio Gallese, and Maria Alessandra Umiltà, "'Embodied Body Language': An Electrical Neuroimaging Study with Emotional Faces and Bodies," *Scientific Reports* 7 (2017): https://doi.org/10.1038/s41598-017-07262-0.
46. Fedorova, "Neurointerfaces," 102.
47. Beaulieu, "Images Are Not the (Only) Truth," 60.
48. "Overview," Visible Human Project, National Library of Medicine, accessed June 10, 2021, https://www.nlm.nih.gov/research/visible/visible_human.html. See Maaike Bleeker, *Visuality in the Theatre: The Locus of Looking* (London: Palgrave Macmillan, 2008), 147–59.
49. Russell A. Poldrack, "Can Cognitive Processes Be Inferred from Neuroimaging Data?," *Trends in Cognitive Sciences* 10, no. 2 (2006): 59–63.
50. Giandomenico Iannetti, email message to the author, February 3, 2020.
51. Giandomenico Iannetti, Tim V. Salomons, Massieh Moayedi, André Mouraux, and Karen D. Davis, "Beyond Metaphor: Contrasting Mechanism of Social and Physical Pain," *Trends in Cognitive Sciences* 17, no. 8 (2013): 374.
52. Fedorova, "Neurointerfaces," 95.
53. Vittorio Gallese, "Bodily Selves in Relation: Embodied Simulation as Second-Person Perspective on Intersubjectivity," *Philosophical Transactions of the Royal Society of London B* 369 (2014): http://dx.doi.org/10.1098/rstb.2013.0177.
54. Joseph Dumit, *Picturing Personhood: Brain Scans and Biomedical Identity* (Princeton, NJ: Princeton University Press, 2004), 112.
55. Beaulieu, "Images Are Not the (Only) Truth," 66.
56. Latour, "What Is Iconoclash?," 18.
57. Jan De Vos, "What Is Critique in the Era of the Neurosciences?," in *Neuroscience and Critique: Exploring the Limits of the Neurological Turn*, ed. Jan De Vos and Ed Pluth (New York: Routledge, 2016), 25.
58. Vittorio Gallese and Michele Guerra, *The Empathic Screen: Cinema and Neuroscience*, trans. Frances Anderson (Oxford: Oxford University Press, 2019).
59. Prasad, "Making Images/Making Bodies," 302.
60. Prasad, "Making Images/Making Bodies," 298.
61. Dumit, *Picturing Personhood*, 168.

62. Casini, "Magnetic Resonance Imaging (MRI) as Mirror and Portrait," 87.
63. Maurice Merleau-Ponty, "Eye and Mind," [1964] in *The Primacy of Perception*, ed. James M. Edie, trans. Carleton Dallery (Evanston, IL: Northwestern University Press, 1964), 160.
64. Roger Fry, *Reflections on British Painting* (London: Faber and Faber, 1934), 134–35.
65. Nelson Goodman, *The Languages of Art* (Indianapolis: Hackett, 1968), 3.
66. Goodman, *The Languages of Art*, 25.
67. Ernst Hans Gombrich, *Art and Illusion: A Study in the Psychology of Pictorial Representation* (Princeton, NJ: Princeton University Press, 1960), 14.
68. Barbara Maria Stafford, *Body Criticism: Imaging the Unseen in Enlightenment Art and Medicine* (Cambridge, MA: MIT Press, 1991), xix.
69. Jan De Vos, "The Iconographic Brain: A Critical Philosophical Inquiry Into (the Resistance of) the Image," *Frontiers in Human Neuroscience* 8 (2014), http://doi.org/10.3389/fnhum.2014.00300.
70. Max R. Bennett and Peter Michael Stephan Hacker, *Philosophical Foundation of Neuroscience* (Oxford: Blackwell, 2003), chap. 14.4, 405.
71. W. J. T. Mitchell, "Realism and the Digital Image," in *Travels in Intermedia: ReBlurring the Boundaries*, ed. Bernd Herzogenrath (Hanover, NH: Dartmouth College Press, 2012), 49.
72. Nicholas Wade, "Perceptual Portraits," in *The Neurobiology of Painting*, ed. Frank Clifford Rose (San Diego: Elsevier, 2006), 17–38.
73. Simona Morini and Maria Perosino, eds., *Specchi: Scienza e coscienza allo specchio* (Turin: Codice, 2006), exhibition catalogue, 9.
74. Siân Ede, *Art and Science* (London: Tauris, 2005), 135.
75. Franko B, *I'm Not Your Babe*, Franko B: Personal/Political/Poetic, accessed June 10, 2021, http://www.franko-b.com/Im_Not_Your_Babe.html.
76. Ede, *Art and Science*, 145.
77. Morten Søndergaard and Peter Weibel, eds., *Thorbjørn Lausten's Visual Systems: Magnet* (Roskilde: MSFK—Museum of Contemporary Art, Roskilde, 2008), exhibition catalogue, 29.
78. Gen. 1:28 (NRSV). See Eduardo Kac, ed., *Biotechnology, Art and Culture* (Cambridge, MA: MIT Press, 2007), accessed June 10, 2021, www.ekac.org/geninfo.html.
79. Barbara Maria Stafford, "From Genetic Perspective to Biohistory: The Ambiguities of Looking Down, Across, and Beyond," in *Sign of Life: Bio Art and Beyond*, ed. Eduardo Kac (Cambridge, MA: MIT Press, 2007), 377.
80. David Edwards, *Artscience: Creativity in the Post-Google Generation* (Cambridge, MA: Harvard University Press, 2009), 20.
81. Philippe Descola, "La double vie des images," in *Penser l'image II: Anthropologies du visuel*, ed. Emmanuel Alloa (Paris: Les Presses du Réel, 2015), 131–45.
82. Fernando Vidal, "Desire, Indefinite Lifespan, and Transgenerational Brains in Literature and Film," *Theory & Psychology* 26, no. 5 (2016): 665–80.
83. Helen Chadwick, *Self-Portrait*, National Galleries of Scotland, accessed July 18, 2021, https://www.nationalgalleries.org/art-and-artists/42351/self-portrait.

84. Joyce Cutler Shaw, "'The Anatomy Lesson': The Body, Technology and Empathy," *Leonardo* 27, no. 1 (1994): 29–38.
85. Emma Taggart, "Psychiatrist Hand-Knits an Anatomically Correct Replica of the Human Brain," *My Modern Met*, December 16, 2018, accessed July 21, 2020, https://mymodernmet.com/knitted-human-brain-dr-karen-norberg/.
86. Susan Aldworth, "Cogito Ergo Sum 1, 2 & 3: Chasing Consciousness," accessed June 10, 2021, https://susanaldworth.com/works/cogito-ergo-sum-1-2-3/.
87. Constance Jacobson, "Constance Jacobson: Prints and Works on Paper," Archives, accessed June 10, 2021, https://www.constancejacobson.com/archives/brains/view/1148247/1/1165865.
88. Marta de Menezes, *Functional Portraits*, accessed June 10, 2021, https://martademenezes.com/art/portraits/functional-portraits/.
89. Angela Palmer, *Brain of the Artist*, National Galleries of Scotland, accessed June 10, 2021, https://www.nationalgalleries.org/art-and-artists/129231/brain-artist.
90. Amanpreet Badhwar and Estrid Jakobsen, "The Interplay Between Neurocience and Art," Organization for Human Brain Mapping, updated June 3, 2017, accessed July 21, 2020, https://www.ohbmbrainmappingblog.com/blog/the-interplay-between-neuroscience-and-art.
91. Laura Jacobson, *Resonance Punctuated*, accessed July 16, 2021, http://laurajacobson.com/brain-scapes/0c9l33t7sd8hi4nma6ebwfkm83b7qn.
92. "Pierre Huyghe: Of Ideal," Hauser & Wirth, online exhibition, September 21, 2020, accessed July 20, 2021, https://www.hauserwirth.com/hauser-wirth-exhibitions/30409-pierre-huygheof-ideal.
93. Barbara Maria Stafford and Horst Bredekamp, "One Step Beyond Hyperrealism," interview/discussion, *Tate Etc.* 6 (January 2006), accessed June 11, 2021, https://www.tate.org.uk/tate-etc/issue-5-autumn-2005/one-step-beyond.
94. John Michel Krois, "Enactivism and Embodiment in Picture Acts," in *Sehen und Handeln*, ed. Horst Bredekamp and John Michel Krois (Berlin: Akademie Verlag, 2011), 17.
95. Fedorova, "Neurointerfaces," 105.
96. Aud Sissel Aurora Hoel, "Images as Active Powers for Reality: A Simondonian Approach to Medical Imaging," in *Dynamis of the Image*, ed. Emmanuel Alloa and Chiara Cappelletto (Berlin: De Gruyter, 2020), 310.
97. De Vos, "The Iconographic Brain."
98. Andrew Carnie, *Magic Forest: Slide Dissolve Work*, accessed June 10, 2021, www.tram.ndo.co.uk/magicforest.htm.
99. "Mariko Mori's UFO—Your Thoughts Come to Life," Public Delivery, accessed June 10, 2021, https://publicdelivery.org/mariko-mori-wave-ufo/.
100. Greg Dunn, *Self Reflected*, Greg Dunn Neuro Art, accessed June 10, 2021, https://www.gregadunn.com/self-reflected/.
101. Laura Jade, *Brainlight*, accessed June 10, 2021, https://laurajade.com.au/interactive-brain-light-research-project/.
102. Fedorova, "Neurointerfaces," 103.
103. Martin Kemp, "From Science in Art to the Art of Science," *Nature* 434 (March 17, 2005): 309.
104. Gombrich, *Art and Illusion*, 176.

## 6. BRAINS ON STAGE

1. Philostratus, *The Life of Apollonius of Tyana: Volume I; Books 1–5*, trans. Frederick C. Conybeare, Loeb Classical Library 16 (1912; repr., Cambridge, MA: Harvard University Press, 2000), bk. 2.22, 177.
2. Tim Ingold, *Making: Anthropology, Archaeology, Art and Architecture* (New York: Routledge, 2013).
3. Vittorio Gallese and Alessandro Gattara, "Embodied Simulation, Aesthetics, and Architecture: An Experimental Aesthetic Approach," in *Mind in Architecture: Neuroscience, Embodiment, and the Future of Design*, ed. Sarah Robinson and Juhani Pallasmaa (Cambridge, MA: MIT Press, 2015), 3920 of 6575, Kindle.
4. Beatrice Sbriscia-Fioretti, Cristina Berchio, David Freedberg, Vittorio Gallese, and Maria A. Umiltà, "ERP Modulation During Observation of Abstract Paintings by Franz Kline," *PLOS One* 8, no. 10 (2013): https://doi.org/10.1371/journal.pone.0075241.
5. Robert Vischer, "On the Optical Sense of Form: A Contribution to Aesthetics" [1873], in *Empathy, Form, and Space: Problems in German Aesthetics, 1873–1893*, trans. and ed. Harry Francis Mallgrave and Eleftherios Ikonomou (Santa Monica, CA: Getty Center for the History of Art and the Humanities, 1993), 94.
6. Adolf Hildebrand, "The Problem of Form in the Fine Arts" [1893], in Mallgrave and Ikonomou, *Empathy, Form, and Space*, 229.
7. Giacomo Rizzolatti and Corrado Sinigaglia, *Mirrors in the Brain: How Our Minds Share Actions, Emotions and Experience*, trans. Frances Anderson (Oxford: Oxford University Press, [2006] 2008), 50.
8. John Dewey, "Affective Thought," in *Later Works*, ed. A. J. Boydstone, vol. 2 (Carbondale-Edwardsville: Southern Illinois University Press, [1926] 1988). Italics mine.
9. George E. Hein, "John Dewey and Albert C. Barnes: A Deep and Mutually Rewarding Friendship," *Dewey Studies* 1, no. 1 (Spring 2017): 44–78.
10. Albert Barnes, *The Art in Painting*, 3rd ed. (Merion, PA: Barnes Foundation, [1925] 1937), 6–7.
11. Henry Moore, "The Sculptor Speaks," in *Henry Moore on Sculpture* (London: Macdonald, 1966), 62.
12. Jean d'Udine, *L'art et le geste* (Paris: F. Alcan, 1910), 63. Translated by Samuel Fleck.
13. Jean d'Udine, *De la corrélation des sons et des couleurs en art* (Paris, 1897).
14. Rizzolatti and Sinigaglia, *Mirrors in the Brain*, ix.
15. Vittorio Gallese, "Il corpo teatrale: mimetismo, neuroni specchio, simulazione incarnata," *Culture teatrali* 16 (2008): 14.
16. Corinne Jola, Shantel Ehrenberg, and Dee Reynolds, "The Experience of Watching Dance: Phenomenological-Neuroscience Duets," *Phenomenology and the Cognitive Sciences* 11 (2012): 17–37.
17. David Freedberg, "From Absorption to Judgment: Empathy in Aesthetic Response," in *Empathy: Epistemic Problems and Cultural-Historical Perspectives of a Cross-Disciplinary Concept*, ed. Vanessa Lux and Sigrid Weigel (London: Palgrave MacMillan, 2017), 155.

18. Juhani Pallasmaa, *The Eyes of the Skin: Architecture and the Senses* (Chichester, UK: Wiley, [1996] 2012), 41.
19. Juhani Pallasmaa, "Body, Mind, and Imagination: The Mental Essence of Architecture," in Robinson and Pallasmaa, *Mind in Architecture*, loc. 51 of 1279, Kindle.
20. Vittorio Gallese, "A Bodily Take on Aesthetics: Performativity and Embodied Simulation," in *The Extended Theory of Cognitive Creativity: Interdisciplinary Approaches to Performativity*, ed. Antonio Pennisi and Alessandra Falzone (Cham, Switzerland: Springer, 2020), 139.
21. Victoria Pitts-Taylor, *The Brain's Body: Neuroscience and Corporeal Politics* (Durham, NC: Duke University Press, 2016), 45.
22. Giuseppe Di Pellegrino, Luciano Fadiga, Leonardo Fogassi, Vittorio Gallese, and Giacomo Rizzolatti, "Understanding Motor Events: A Neurophysiological Study," *Experimental Brain Research* 91 (1992): 176–80; Vittorio Gallese, Luciano Fadiga, Leonardo Fogassi, and Giacomo Rizzolatti, "Action Recognition in the Premotor Cortex," *Brain* 119 (1996): 593–609.
23. Vittorio Gallese, "Mirroring, a Liberated Embodied Simulation and Aesthetic Experience," in *Mirror Images: Reflections in Art and Medicine* (Thun, Switzerland: Verlag für moderne Kunst, Kunstmuseum Thun, 2017), 32.
24. Fotini Vassiliou, "Aesthetic Disinterestedness in Neuroaesthetics: A Phenomenological Critique," *Aesthetic Investigations* 4, no. 1 (2020): 79.
25. Marco Iacoboni, *I neuroni specchio: Come capiamo ciò che fanno gli altri* (Turin: Bollati Boringhieri, 2008), 17.
26. Vilayanur S. Ramachandran, *Mirror Neurons and Imitation Learning as the Driving Force Behind "The Great Leap Forward" in Human Evolution*, accessed February 11, 2020, https://www.edge.org/3rd_culture/ramachandran/ramachandran_index.html.
27. Vittorio Gallese, "The 'Shared Manifold' Hypothesis: From Mirror Neurons to Empathy," *Journal of Consciousness Studies* 8, no. 5–7 (2001): 33–50; and Vittorio Gallese, "'Being Like Me': Self-Other Identity, Mirror Neurons and Empathy," in *Perspectives on Imitation: From Cognitive Neuroscience to Social Science*, ed. Susan Hurley and Nick Chater, 2 vols. (Cambridge, MA: MIT Press, 2005).
28. Bernard Berenson, *Aesthetics and History in the Visual Arts* (New York: Pantheon, 1948), 69–71.
29. Pier Francesco Ferrari, Vittorio Gallese, Giacomo Rizzolatti, and Leonardo Fogassi, "Mirror Neurons Responding to the Observation of Ingestive and Communicative Mouth Actions in the Monkey Ventral Premotor Cortex," *European Journal of Neuroscience* 17 (2003): 1703–14.
30. Leonardo Fogassi and Vittorio Gallese, "The Neural Correlates of Action Understanding in Non-Human Primates," in *Mirror Neurons and the Evolution of Brain and Language*, ed. Maxim I. Stamenov and Vittorio Gallese (Amsterdam: John Benjamins, 2002), 13–15.
31. Maria Alessandra Umiltà, Evelyne Kohler, Vittorio Gallese, Leonardo Fogassi, Luciano Fadiga, Christian Keysers, and Giacomo Rizzolatti, "'I Know What You Are Doing': A Neurophysiological Study," *Neuron* 31 (2001): 155–65.

32. Leonardo Fogassi, Pier Francesco Ferrari, Benno Gesierich, Stefano Rozzi, Fabian Chersi, and Giacomo Rizzolatti, "Parietal Lobe: From Action Organization to Intention Understanding," *Science* 308 (2005): 662–67.
33. Giacomo Rizzolatti, Luciano Fadiga, Massimo Matelli, Valentino Bettinardi, Eraldo Paulesu, Daniela Perani, and Ferrucio Fazio, "Localization of Grasp Representations in Humans by PET: 1. Observation Versus Execution," *Experimental Brain Research* 111 (1996): 246–52; Scott T. Grafton, Michael A. Arbib, Luciano Fadiga, and Giacomo Rizzolatti, "Localization of Grasp Representations in Humans by Positron Emission Tomography: 2. Observation Compared with Imagination," *Experimental Brain Research* 112 (1996): 103–11.
34. Andrew N. Meltzoff and M. Keith Moore, "Explaining Facial Imitation: A Theoretical Model," *Early Development and Parenting* 6 (1997): 179–92.
35. Pierre Jacob and Marc Jeannerod, *Ways of Seeing: The Scope and Limits of Visual Cognition* (Oxford: Oxford University Press, 2003), 230–34.
36. Giovanni Buccino, Fausta Lui, Nicola Canessa, Ilaria Patteri, Giovanna Lagravinese, Francesca Benuzzi, Carlo A. Porro, and Giacomo Rizzolatti, "Neural Circuits Involved in the Recognition of Actions Performed by Nonconspecifics: An fMRI Study," *Journal of Cognitive Neuroscience* 16 (2004): 114–26.
37. Corinne Jola, Angharad Davis, and Patrick Haggard, "Proprioceptive Integration and Body Representation: Insights Into Dancers' Expertise," *Experimental Brain Research* 213, no. 2–3 (2011): 257–65.
38. Paul B. Preciado, *Counter-Sexual Manifesto* [2000], trans. Kevin Gerry Dunn (New York: Columbia University Press, 2018), 11.
39. Etienne Decroux, *Words on Mime*, trans. Mark Piper (Claremont, CA: Mime Journal, [1963] 1985), 85.
40. Marc Jeannerod, "The Representing Brain: Neural Correlates of Motor Intention and Imagery," *Behavioral Brain Sciences* 17 (1997): 187–245.
41. Rizzolatti and Sinigaglia, *Mirrors in the Brain*, 106.
42. Ivar Hagendoorn, "Some Speculative Hypotheses About the Nature and Perception of Dance and Choreography," *Journal of Consciousness Studies* 11, no. 3–4 (2004): 80.
43. Beatriz Calvo-Merino, Daniel E. Glaser, Julie Grèzes, Richard E. Passingham, and Patrick Haggard, "Action Observation and Acquired Motor Skills: An fMRI Study with Expert Dancers," *Cerebral Cortex* 15 (August 2005): 1243, 1248.
44. Beatriz Calvo-Merino, Julie Grèzes, Daniel E. Glaser, Richard E. Passingham, and Patrick Haggard, "Seeing or Doing? Influence of Visual and Motor Familiarity in Action Observation," *Current Biology* 16 (2006): 1905–10.
45. Ivar Hagendoorn, "Cognitive Dance Improvisation: How Study of the Motor System Can Inspire Dance (and Vice Versa)," in *Leonardo* 36, no. 3 (2003): 221.
46. Marc Jeannerod, *La nature de l'esprit: Science cognitive et cerveau* (Paris: Odile Jacob, 2002), 47.
47. Marco Iacoboni, Istvan Molnar-Szakacs, Vittorio Gallese, Giovanni Buccino, John Mazziotta, and Giacomo Rizzolatti, "Grasping the Intentions of Others with One's Own Mirror Neuron System," *PLOS Biology* 3 (2005): 530, 533.

48. Di Pellegrino, Fadiga, Fogassi, Gallese, and Rizzolatti, "Understanding Motor Events," 179.
49. Kimberly L. Kontson, Murad Megjhani, Justin A. Brantley, Jesus Cruz-Garza, Sho Nakagome, Dario Robleto, Michelle White, et al., "Your Brain on Art: Emergent Cortical Dynamics During Aesthetic Experiences," *Frontiers in Human Neuroscience* 9 (2015): https://doi.org/10.3389/fnhum.2015.00626.
50. Such experiments often use electromagnetically isolated Faraday cages to prevent the signal from being contaminated by network noise: the subject performs the task in the cabin and the experimenter's control station is outside. Compared with other brain-imaging techniques, however, EEG facilitates recording in more ecological contexts, in interactions with other people, as well as in studies conducted outside the laboratory.
51. Gabriele Sofia, "The Spectator's 'Making Sense': An Interdisciplinary Research Between Theatre and Neurosciences," in *Making Sense: Merging Theory and Praxis*, ed. Bandy Lee and Lorna Collins (Bern: Peter Lang, 2013), 175–88.
52. Personal communication to the author during the rehearsals for a staging of Shakespeare's *Othello*, on which the author served as theater company assistant, at the Venice Biennale Theater, summer 1999.
53. Jean-Baptiste Du Bos, *Critical Reflections on Poetry, Painting and Music*, ed. James O. Young and Margaret Cameron (London: Brill, [1719] 2021) 3.
54. Du Bos, *Critical Reflections*, 3:28.
55. Aristotle, *Poetics*, trans. Anthony Kenny (Oxford: Oxford University Press, 2013), 50a:16–17, 48b:5–9.
56. Barbara Maria Stafford, *Echo Objects: The Cognitive Work of Images* (Chicago: University of Chicago Press, 2007), 77.
57. Vittorio Gallese, personal communication to the author at the international conference "'As-If'—Figures of Imagination, Simulation, and Transposition in the Relation to the Self, Others, and Arts," organized by Sigrid Weigel and Andrea Pinotti, Villa Vigoni, Como, Italy, June 3–6, 2014.
58. Alva Noë, *Varieties of Presence* (Cambridge, MA: Harvard University Press, 2012).
59. Andy Clark, "Re-Inventing Ourselves: The Plasticity of Embodiment, Sensing, and Mind," *Journal of Medicine and Philosophy: A Forum for Bioethics and Philosophy of Medicine* 32, no. 3 (2007): 263–82.
60. Maria A. Umiltà, Ludovic Escola, Irakli Intskirveli, Franck Grammont, Magali Rochat, Fausto Caruana, Ahmad Jezzini, Vittorio Gallese, and Giacomo Rizzolatti, "When Pliers Become Fingers in the Monkey Motor System," *PNAS* 105, no. 6 (2008): 2210.
61. Daria Lippi, Corinne Jola, Victor Jacono, and Gabriele Sofia, "Steps Towards the Art of Placing Science in the Acting Practice: A Performance-Neuroscience Perspective," in *Aesthetics and Neuroscience: Scientific and Artistic Perspective*, ed. Zoï Kapoula and Marine Vernet (Cham, Switzerland: Springer, 2016), 147–48.
62. Chiara Cappelletto, "L'attore, una figura chiasmatica," in *Filosofie sull'attore*, ed. Katia L. Angioletti (Milan: LED, 2010), 63–78.
63. Bertolt Brecht, "Alienation Effects in Chinese Acting," in *Brecht on Theater: The Development of an Aesthetic*, trans. and ed. John Willett, 91–99 (New York: Methuen, [1936] 1966).

64. Helmuth Plessner, "Zur Anthropologie des Schauspielers," [1948], in *Gesammelte Schriften*, ed. Günter Dux, Odo Marquard, and Elisabeth Ströker (Frankfurt: Suhrkamp, 1982), 399–418.
65. Plessner, "Zur Anthropologie des Schauspielers," 399–418. Italics mine.
66. Jean-Pierre Changeux, *Raison et Plaisir* (Paris: Odile Jacob, [1989] 1994), loc. 1432 of 2701, Kindle.
67. David Freedberg and Vittorio Gallese, "Motion, Emotion and Empathy in Esthetic Experience," *Trends in Cognitive Sciences* 11, no. 5 (May 2007): 197–203.
68. Freedberg and Gallese, "Motion, Emotion and Empathy," 202.
69. Cinzia Di Dio, Martina Ardizzi, and Davide Massaro, "Human, Nature, Dynamism: The Effects of Content and Movement Perception on Brain Activations During the Aesthetic Judgment of Representational Paintings," *Frontiers in Human Neurosciences* 9 (2016): https://doi.org/10.3389/fnhum.2015.00705.
70. Melissa M. Littlefield and Jenell Johnson, introduction to *The Neuroscientific Turn: Transdisciplinarity in the Age of the Brain* (Ann Arbor: University of Michigan Press, 2012), 15.
71. David Freedberg, "Empathy, Motion and Emotion," in *Wie sich Gefühle Ausdruck verschaffen: Emotionen in Nahsicht*, ed. Klaus Herding and Antje Krause Wahl (Berlin: Driesen, 2007), 26–27.
72. Maria A. Umiltà, Cristina Berchio, Mariateresa Sestito, David Freedberg, and Vittorio Gallese, "Abstract Art and Cortical Motor Activation: An EEG Study," *Frontiers in Human Neuroscience* 6 (2012): https://doi.org/10.3389/fnhum.2012.00311.
73. Immanuel Kant, *Critique of Judgment*, ed. Nicholas Walker, trans. James Creed Meredith, rev. ed. (Oxford: Oxford University Press, [1790] 2009); Fotini Vassiliou, "Aesthetic Disinterestedness in Neuroaesthetics: A Phenomenological Critique," *Aesthetic Investigations* 4, no. 1 (2020): 77–95.
74. Freedberg, "From Absorption to Judgment," 166.
75. Freedberg, "From Absorption to Judgment," 150.
76. Ludovico Mineo, Alexander Fetterman, Carmen Concerto, Michael Warren, Carmenrita Infortuna, David Freedberg, Eileen Chusid, Eugenio Aguglia, and Fortunato Battaglia, "Motor Facilitation During Observation of Implied Motion: Evidence for a Role of the Left Dorsolateral Prefrontal Cortex," *International Journal of Psychophysiology* 128 (2018): 47–51.
77. Carmen Concerto, Mohomad Al Sawah, Carmenrita Infortuna, David Freedberg, Eileen Chusid, Eugenio Aguglia, and Fortunato Battaglia, "Neural Circuits Underlying Motor Facilitation During Observation of Implied Motion," *Somatosensory and Motor Research* 32, no. 4 (2015): 207–10.
78. Carmen Concerto, Carmenrita Infortuna, Ludovico Mineo, Manuel Pereira, David Freedberg, Eileen Chusid, Eugenio Aguglia, and Fortunato Battaglia, "Observation of Implied Motion in a Work of Art Modulates Cortical Connectivity and Plasticity," *Journal of Exercise Rehabilitation* 12, no. 5 (2016): 417–23.
79. Freedberg, "From Absorption to Judgment," 163.
80. Alfred Gell, *Art and Agency: An Anthropological Theory* (Oxford: Oxford University Press, 1998).

81. Vittorio Gallese, "Brain, Body, Habit and the Performative Quality of Aesthetics," in *Habits: Pragmatist Approaches from Cognitive Neuroscience to Social Science*, ed. Italo Testa and Fausto Caruana (Cambridge: Cambridge University Press, 2021), 388–89.
82. Gallese, "Mirroring, a Liberated Embodied Simulation," 34.
83. Gallese, "Mirroring, a Liberated Embodied Simulation," 33.
84. Gallese, "A Bodily Take on Aesthetics," 141.
85. Vittorio Gallese, "Bodily Selves in Relation: Embodied Simulation as Second-Person Perspective on Intersubjectivity," *Philosophical Transactions of the Royal Society of London B* 369 (2014): http://dx.doi.org/10.1098/rstb.2013.0177.
86. Anna Berti and Francesca Frassinetti, "When Far Becomes Near: Re-Mapping of Space by Tool Use," *Journal of Cognitive Neuroscience* 12 (2000): 415–20.
87. Francesco Ortega and Fernando Vidal, *Being Brains: Making the Cerebral Subject* (New York: Fordham University Press, 2017), 128.
88. Olafur Eliasson, *Contact, 2014*, accessed July 6, 2021, https://olafureliasson.net/archive/artwork/WEK109128/contact.
89. "About," *Roden Crater*, accessed June 21, 2021, https://rodencrater.com/about/.
90. Stefan Kaegi and Jörg Karrenbauer, *Remote X*, Rimini Protokoll, accessed July 16, 2021, https://www.rimini-protokoll.de/website/en/project/remote-x.
91. Jody Oberfelder, "Does the Body Think, Do Your Neurons Dance," *New York Times*, June 27, 2017, https://www.nytimes.com/2017/06/27/arts/dance/does-the-body-think-do-your-neurons-dance-jody-oberfelder-new-york-live-arts.html.
92. Johannes Volkelt, *System der Ästhetik*, vol. 1 (Munich: Beck, 1905), 215–17.
93. Eugenio Barba, *Theatre: Solitude, Craft, Revolt*, trans. Judy Barba (Aberystwyth, UK: Back Mountain, [1996] 1999), 248.
94. Freedberg, "From Absorption to Judgment," 144.

# BIBLIOGRAPHY

## THE HYDRA OF LERNA: ON A BIBLIOGRAPHIC CORPUS

In the past decade the term *neuroaesthetics* has gone from an outlandish neologism to an area of knowledge and a keyword in major catalogues worldwide. The field has seen a tremendous increase in contributions. Everything from manifestos to experimental research papers on artistic practices, spectatorship, and related theoretical issues is now available, via books and journals, to experts from different backgrounds as well as to lay readers.

Yet the growing number of writings does not correspond to a significant variety of authors. The neuroscientists involved in the neuroaesthetic enterprise make up a relatively small family whose members are spread throughout the globe and who find themselves using the same journals, databases, and language—English, broken though it may be. They often address and expand on similar arguments across multiple texts. As for contributions by philosophers and by those working in the humanities more broadly, the state of research lacks input from non-Anglophone scholars and traditions. Taken together, therefore, these studies suffer from a considerable blind spot that requires illuminating.

The task of doing so exposes the limitations of the usual strategies of bibliographic development (updating and expanding). In tackling it,

I have been struck by the emergence of an awkward disclaimer: "To my knowledge, the references to this matter are the following." Have references not always been a question of the author's knowledge? What could be driving this *excusatio non petita* if not the unease, spreading among scholars in tough interdisciplinary arenas, at trying to keep pace with a rapidly expanding bibliography fed by publish-or-perish policies, which too often encourage redundant and hyperspecialized contributions (primarily to secure public attention and funding), turn knowledge production into a kind of treadmill, and neglect more critical and genealogical viewpoints?

To manage such a corpus of literature, I appealed to the following criteria:

1. Methodology and media: Consistent with Elkins's "problem of the pencil," mentioned in chapter 1, sources are classified as either "top-down and argument-based," where grounded in neat argumentation, or "bottom-up and experiment-based," where grounded in empirical research and procedures and in statistical methods. This choice does not imply that all authors in the first group are humanists and those in the second are practitioners in the lab; indeed, it is worth underscoring the extent to which scientists working in the neuroaesthetic field have authored opinion papers and engaged in philosophical debates. This comes as no surprise. As Pitts-Taylor makes clear in *The Brain's Body*, once the brain is conceived of as a "social problem," as the biological ground for the Self and for social life, neuroscientists take the floor as philosophers and sociologists.

2. Subject: In both cases, sources are further subdivided based on their approach to neuroaesthetic questions—whether strictly theoretical, tied to visual artifacts, or related to performances.

3. Field: To address the lack of variety and critical-historical awareness in the respective scholarly contributions, notably from the philosophical and the artistic fields, a third, complementary section is included.

It bears noting that studies dealing with the correlation between neurobiological mechanisms and rewarding processes, taste and disgust in everyday contexts, attentional and affective discrimination tasks, and so on, are generally and consciously excluded, for—pace Pinker—I do not consider artworks and things like cheesecake to be suitable items for comparison.

# TOP-DOWN AND ARGUMENT-BASED NEUROSCIENTIFIC SOURCES

## AESTHETICS

Adli, Mazda, Maximilian Berger, Eva-Lotta Brakemeier, Ludwig Engel, Joerg Fingerhut, Ana Gomez-Carillo, Rainer Hehl, et al. "Neurourbanistik—ein methodischer Schulterschluss zwischen Stadtplanung und Neurowissenschaften." *Die Psychiatrie* 13, no. 2 (2016): 70–78.

Agnati, Luigi Francesco, Achille Agnati, Francisco Mora, and Kjell Fuxe. "Does the Human Brain Have Unique Genetically Determined Networks Coding Logical and Ethical Principles and Aesthetics? From Plato to Novel Mirror Networks." *Brain Research Reviews* 55 (2007): 68–77.

American Psychological Association. *Functional Magnetic Resonance Imaging: A New Research Tool.* Washington, DC: American Psychological Association, 2007. Accessed July 14, 2021. https://www.yumpu.com/en/document/read/10408039/functional-magnetic-resonance-imaging-a-new-research-tool.

Ball, Philip. "Neuroaesthetics Is Killing Your Soul: Can Brain Scans Ever Tell Us Why We Like Art?" *Nature*, March 22, 2013. https://doi.org/10.1038/nature.2013.12640.

Bao, Yan, and Ernst Pöppel. "Anthropological Universal and Cultural Specifics: Conceptual and Methodological Challenges in Cultural Neuroscience." *Neuroscience & Biobehavioral Reviews* 36, no. 9 (2012): 2143–46.

Bara, Ionela, Richard J. Binney, Rob Ward, and Richard Ramsey. "A Generalised Semantic Cognition Account of Aesthetic Experience." *PsyArXiv preprint*. December 22, 2020. http://doi.org/10.31234/osf.io/wbeqx.

Barcaro, Umberto, and Marco Paoli. "Dreaming and Neuroesthetics." *Frontiers in Human Neuroscience* 9 (2015). https://doi.org/10.3389/fnhum.2015.00348.

Bassiri, Nima. "Who Are We, Then, If We Are Indeed Our Brains? Reconsidering a Critical Approach to Neuroscience." In De Vos and Pluth, *Neuroscience and Critique*, 41–61.

Beaulieu, Anne. "Voxels in the Brain: Neuroscience, Informatics and Changing Notions of Objectivity." *Social Studies of Science* 31, no. 5 (October 2001): 635–80.

Breidbach, Olaf. *Das Anschauliche oder über die Anschauung von Welt: ein Beitrag zur Neuronalen Ästhetik*. Vienna: Springer, 2000.

——. "The Beauties and the Beautiful: Some Considerations from the Perspective of Neuronal Aesthetics." In Voland and Grammar, *Evolutionary Aesthetics*, 39–68.

——. "Die Innenwelt der Außenwelt—Weltkonstitution im Hirngewebe? Zur Konturierung einer neuronalen Ästhetik." In Breidbach and Clausberg, *Video ergo sum*, 34–60.

Breidbach, Olaf, and Karl Clausberg, eds. *Video ergo sum: Repräsentation nach innen und außen zwischen Kunst- und Neurowissenschaften*. Hamburg: Hans-Bredow-Institut, 1999.

Brielmann, Aenne A., and Denis G. Pelli. "Aesthetics." *Current Biology* 28 (2018): 859–63.

Brown, Steven. "Toward a Unification of the Arts." *Frontiers in Psychology* 9 (2018). https://doi.org/10.3389/fpsyg.2018.01938.

Brown, Steven, and Ellen Dissanayake. "The Arts Are More Than Aesthetics: Neuroaesthetics as Narrow Aesthetics." In Skov and Vartanian, *Neuroaesthetics*, 43–57.

Bullot, Nicolas J. "History, Causal Information, and the Neuroscience of Art: Toward a Psycho-Historical Theory." In Wolfe, *Brain Theory*, 241–63.

Bundgaard, Peer F. "Feeling, Meaning, and Intentionality—A Critique of the Neuroaesthetics of Beauty." *Phenomenology and the Cognitive Sciences* 14 (2015): 781–801.

Carbon, Claus-Christian. "Empirical Aesthetics: In Quest of a Clear Terminology and Valid Methodology." In *Exploring Transdisciplinarity in Art and Sciences*, ed. Zoï Kapoula, Emmanuelle Volle, Julien Renoult, and Moreno Andreatta, 107–19. Cham, Switzerland: Springer, 2018.

Carew, Thomas J., and Ramaswami Mani. "The Neurohumanities: An Emerging Partnership for Exploring the Human Experience." *Neuron* 108 (2020): 590–93.

Carroll, Noel, Margaret Moore, and William P. Seeley. "The Philosophy of Art and Aesthetics, Psychology and Neuroscience: Studies in Literature, Music and Visual Art." In Shimamura and Palmer, *Aesthetics Science*, 31–62.

Cattaneo, Zaira. "Noninvasive Brain Stimulation: An Overview of Available Approaches for Research in Neuroaesthetics." *Empirical Studies of the Arts* 37, no. 2 (2019): 153–71.

Cela-Conde, Camilo J., and Francisco J. Ayala. "Art and Brain Co-evolution." In Huston, Nadal, Mora, Agnati, and Cela-Conde, *Art, Aesthetics and the Brain*, 408–25.

Chancellor, Bree, and Anjan Chatterjee. "Brain Branding: When Neuroscience and Commerce Collide." *AJOB Neuroscience* 2, no. 4 (2011): 18–27.

Changeux, Jean-Pierre. "Art and Neuroscience." *Leonardo* 27, no. 3 (1994): 189–201.

——. "Beauty in the Brain: For a Neuroscience of Art." *Rendiconti Lincei* 23 (2012): 315–20.

——. *The Good, the True and the Beautiful: A Neuronal Approach*. Trans. and rev. Laurence Garey. New Haven, CT: Yale University Press, [2008] 2012.

——. *Raison et Plaisir*. Paris: Odile Jacob, [1989] 1994. Kindle.

Changeux, Jean-Pierre, and Paul Ricoeur. *What Makes Us Think? A Neuroscientist and a Philosopher Argue About Ethics, Human Nature and the Brain*. Trans. M. B. DeBevoise. Princeton, NJ: Princeton University Press, [1998] 2000.

Chatterjee, Anjan. *The Aesthetic Brain: How We Evolved to Desire Beauty and Enjoy Art*. Oxford: Oxford University Press, 2014.

——. "Neuroaesthetics." In *The Oxford Encyclopedia of Aesthetics*, ed. Michael Kelly, 2nd ed., 481–91. New York: Oxford University Press, 2014.

——. "Neuroaesthetics: A Coming of Age Story." *Journal of Cognitive Neuroscience* 23, no. 1 (2011): 53–62.

——. "Neuroaesthetics: Growing Pains of a New Discipline." In Shimamura and Palmer, *Aesthetics Science*, 299–317.

——. "Scientific Aesthetics: Three Steps Forward." *British Journal of Psychology* 105 (2014): 465–67.

——. "Where There Be Dragons: Finding the Edges of Neuroaesthetics." *American Society for Aesthetics Newsletter* 31 (2011): 4–6.

Chatterjee, Anjan, and Oshin Vartanian. "Neuroaesthetics." *Trends in Cognitive Sciences* 18, no. 7 (2014): 370–75.

——. "Neuroscience of Aesthetics." *Annals of the New York Academy of Sciences* 1369 (2016): 172–94.

Chatterjee, Anjan, Page Widick, Rebecca Sternschein, William B. Smith, and Bianca Bromberger. "The Assessment of Art Attributes." *Empirical Studies of the Arts* 28, no. 2 (2010): 207–22.

Christensen, Julia F. "Pleasure Junkies All Around! Why It Matters and Why 'The Arts' Might Be the Answer: A Biopsychological Perspective." *Proceedings of the Royal Society B* 284 (2017): 1–8.

Clausberg, Karl. *Neuronale Kunstgeschichte. Selbstdarstellung als Gestaltungsprinzip*. Vienna: Springer, 1999.

Coburn, Alex, Oshin Vartanian, and Anjan Chatterjee. "Buildings, Beauty, and the Brain: A Neuroscience of Architectural Experience." *Journal of Cognitive Neuroscience* 29, no. 9 (2017): 1521–31.

Consoli, Gianluca. "From Beauty to Knowledge: A New Frame for the Neuropsychological Approach to Aesthetics." *Frontiers in Human Science* 9, no. 290 (2015). https://doi.org/10.3389/fnhum.2015.00290.

Contreras-Vidal, Jose L., Jeannie Kever, Dario Robleto, and James Rosengren. "At the Crossroads of Art and Science: Neuroaesthetics Begins to Come Into Its Own." *Leonardo* 52, no. 1 (2019): 103–6.

Contreras-Vidal, Jose L., Dario Robleto, Jesus G. Cruz-Garza, José M. Azorin, and Chang S. Nam, eds. *Mobile Brain-Body Imaging and the Neuroscience of Art, Innovation and Creativity*. Cham, Switzerland: Springer, 2019.

Conway, Bevil R., and Alexander Rehding. "Neuroaesthetics and the Trouble with Beauty." *PLOS Biology* 11, no. 3 (2013): https://doi.org/10.1371/journal.pbio.1001504.

Croft, James. "The Challenges of Interdisciplinary Epistemology in Neuroaesthetics." *Mind, Brain and Education* 5, no. 1 (2011): 5–11.

Cronan, Todd, ed. "Evaluating Neuroaesthetics." Special issue, *nonsite.org*, no. 2 (2011). Accessed July 21, 2020. https://nonsite.org/issues/issue-2-evaluating-neuroaesthetics/.

Cross, Emily S., and Luca F. Ticini. "Neuroaesthetics and Beyond: New Horizons in Applying the Science of the Brain to the Art of Dance." *Phenomenology and the Cognitive Sciences* 11 (2012): 5–16.

Cubelli, Roberto. "Estetica empirica o neuroestetica? Meglio psicologia dell'arte." *Giornale Italiano di Psicologia* 1 (2020): 123–31.

D'Angelo, Paolo. "La sfida delle neuroscienze." *Rivista internazionale di Filosofia Contemporanea* 6, no. 1 (2018): 179–98.

De Vos, Jan. "What Is Critique in the Era of the Neurosciences?" In De Vos and Pluth, *Neuroscience and Critique*, 22–40.

De Vos, Jan, and Ed Pluth, eds. *Neuroscience and Critique: Exploring the Limits of the Neurological Turn*. New York: Routledge, 2016.

Dresler, Martin, ed. *Neuroästhetik: Kunst—Gehirn Wissenschaft*. Leipzig: Seemann, 2009.

Dumit, Joseph. "The Fragile Unity of Neuroscience." In De Vos and Pluth, *Neuroscience and Critique*, 223–31.

Emery, Alan Eglin Heathcote. "Neurogenetics in Art." In Rose, *The Neurobiology of Painting*, 235–40.

Fayn, Kirill, and Paul J. Silvia. "States, People, and Contexts: Three Psychological Challenges for the Neuroscience of Aesthetics." In Huston, Nadal, Mora, Agnati, and Cela-Conde, *Art, Aesthetics and the Brain*, 40–56.

Fingerhut, Joerg. "Enactive Aesthetics and Neuroaesthetics." *Phenomenology and the Mind* 14 (2018): 80–97.

Fitzpatrick, Susan M. "Functional Brain Imaging: Neuro-Turn or Wrong Turn?" In Littlefield and Johnson, *The Neuroscientific Turn*, 180–99.

Gallese, Vittorio, and George Lakoff. "The Brain's Concepts: The Role of the Sensory-Motor System in Conceptual Knowledge." *Cognitive Neuropsychology* 22 (2005): 455–79.

Gasperoni, Lidia. "Sensible Schemes in Aesthetic Experience: Neuroaesthetics and Transcendental Philosophy Compared." *Aisthesis* 1, no. 1 (2017): 63–73.

Goldberg, Elkhonon, Brent A. Funk, and Kenneth Podell. "How the Brain Deals with Novelty and Ambiguity: Implications for Neuroaesthetics." *Rendiconti Lincei* 23 (2012): 227–38.

Gruber, David R. "Three Forms of Neuro-Realism: Explaining the Persistence of the 'Uncritical Real' in Popular Neuroscience News." *Written Communication* 34, no. 2 (2017): 189–223.

Hadravova, Tereza. "Aesthetic Experts." *ESPES* 8, no. 1 (2019). https://doi.org/10.5281/zenodo.3255297.

Hirstein, William. "Neuroaesthetics: Responding to the Critics." *Psychology Today*, December 18, 2012.

Hobson, J. Allan, and Hellmut Wohl. *Dagli angeli ai neuroni: l'arte e la nuova scienza dei sogni.* Fidenza: Mattioli 1885, [2005] 2007.

Huston, Joseph P., Marcos Nadal, Francisco Mora, Luigi F. Agnati, and Camilo J. Cela-Conde, eds. *Art, Aesthetics and the Brain.* Oxford: Oxford University Press, 2015.

Hutto, Daniel D. "Enactive Aesthetics: Philosophical Reflections on Artful Minds." In Scarinzi, *Aesthetics and the Embodied Mind*, 211–27.

Hutzler, Florian. "Reverse Inference Is Not a Fallacy Per Se: Cognitive Processes Can Be Inferred from Functional Imaging Data." *NeuroImage* 84 (2014): 1061–69.

Hyman, John. "Art and Neuroscience." In *Beyond Mimesis and Convention*, ed. Roman Frigg and Matthew C. Hunter, 245–61. Dordrecht, Netherlands: Springer, 2010.

Iigaya, Kiyohito, John P. O'Doherty, and G. Gabrielle Starr. "Progress and Promise in Neuroaesthetics." *Neuron* 108 (2020): 594–96.

Ishizu, Tomohiro, and Semir Zeki. "Toward a Brain-Based Theory of Beauty." *PLOS One* 6 (2011): https://doi.org/10.1371/journal.pone.0021852.

Kandel, Eric R. *Reductionism in Art and Brain Science: Bridging the Two Cultures.* New York: Columbia University Press, 2016.

Kandel, Eric R., and Sarah Mack. "A Parallel Between Radical Reductionism in Science and Art." *Annals of the New York Academy of Sciences* 1001 (2003): 272–94.

Kapoula, Zoï, and Marine Vernet, eds. *Aesthetics and Neuroscience: Scientific and Artistic Perspective.* Cham, Switzerland: Springer, 2016.

Kawabata, Hideaki, and Semir Zeki. "The Neural Correlates of Beauty." *Journal of Neurophysiology* 91 (2004): 1699–705.

Kirsch, Louise P., Cosimo Urgesi, and Emily S. Cross. "Shaping and Reshaping the Aesthetic Brain: Emerging Perspectives on the Neurobiology of Embodied Aesthetics." *Neuroscience & Biobehavioral Reviews* 62 (2016): 56–68.

Kubovy, Michael. "Neuroaesthetics: Maladies and Remedies." *Art & Perception* 8 (2020): 1–26.

Lakoff, George. "The Neuroscience of Form in Art." In Turner, *The Artful Mind*, 153–69.

Lauring, Jon O., ed. *An Introduction to Neuroaesthetics: The Neuroscientific Approach to Aesthetic Experience, Artistic Creativity, and Arts Appreciation*. Copenhagen: Museum Tusculanum, 2014.

——. "The Theoretical and Methodological Backdrop of Neuroaesthetics." In Lauring, *An Introduction to Neuroaesthetics*, 71–105.

Leder, Helmut. "Next Steps in Neuroaesthetics: Which Processes and Processing Stages to Study?" *Psychology of Aesthetics, Creativity and the Arts* 7 (2013): 27–37.

Leder, Helmut, Gernot Gerger, and David Brieber. "Aesthetics Appreciation: Convergence from Experimental Aesthetics and Physiology." In Huston, Nadal, Mora, Agnati, and Cela-Conde, *Art, Aesthetics and the Brain*, 57–78.

Leder, Helmut, and Marcos Nadal. "Ten Years of a Model of Aesthetic Appreciation and Aesthetic Judgments: The Aesthetic Episode—Developments and Challenges in Empirical Aesthetics." *British Journal of Psychology* 105 (2014): 443–64.

Leder, Helmut, and Pablo P. L. Tinio. "Experimental Aesthetics." In Lauring, *An Introduction to Neuroaesthetics*, 51–65.

Lehrer, Jonah. "Unlocking the Mysteries of The Artistic Mind." *Psychology Today*, July 1, 2010.

Levinson, Jerrold, and Jenefer Robinson, eds. "Art, Mind and Cognitive Science." Special issue, *Journal of Aesthetics and Art Criticism* 62, no. 2 (Spring 2004).

Littlefield, Melissa M., and Jenell Johnson, eds. *The Neuroscientific Turn: Transdisciplinarity in the Age of the Brain*. Ann Arbor: University of Michigan Press, 2012.

Marin, Manuela, M. "Crossing Boundaries: Toward a General Model of Neuroaesthetics." *Frontiers in Human Neuroscience* 9 (2015). https://doi.org/10.3389/fnhum.2015.00443.

Martín-Aragúz, Antonio, José Javier Campos-Bueno, Victor Fernández-Armayor, and Octavio de Juan Ayala. *Neuroestética*. Madrid: Saned, 2010.

Martindale, Colin, Paul Locher, and Vladimir M. Petrov, eds. *Evolutionary and Neurocognitive Approaches to Aesthetics, Creativity and the Arts*. Amityville, NY: Baywood, 2007.

Massey, Edward Wayne. "Neurology and Sherlock Holmes." In Martindale, Locher, and Petrov, *Evolutionary and Neurocognitive Approaches to Aesthetics, Creativity and the Arts*, 357–70.

Massey, Irving. *The Neural Imagination: Aesthetic and Neuroscientific Approaches to the Arts*. Austin: University of Texas Press, 2009.

Mora, Francisco. "Neuroculture: A New Cultural Revolution?" In Huston, Nadal, Mora, Agnati, and Cela-Conde, *Art, Aesthetics and the Brain*, 3–18.

Nadal, Marcos. "Constraining Hypotheses on the Evolution of Art and Aesthetic Appreciation." In Skov and Vartanian, *Neuroaesthetics*, 103–29.

Nadal, Marcos, and Anjan Chatterjee. "Neuroaesthetics and Art's Diversity and Universality." *WIREs Cognitive Science* 10, no. 3 (2018). https://doi.org/10.1002/wcs.1487.

Nadal, Marcos, Antoni Gomila, and Alejandro Gàlvez-Pol. "A History for Neuroaesthetics." In Lauring, *An Introduction to Neuroaesthetics*, 3–49.

Nadal, Marcos, and Marcus T. Pearce. "The Copenhagen Neuroaesthetics Conference: Prospects and Pitfalls for an Emerging Field." *Brain and Cognition* 76 (2011): 172–83.

Nadal, Marcos, and Martin Skov. "Art Is Not Special: An Assault on the Last Lines of Defense Against the Naturalization of the Human Mind." *Reviews in the Neurosciences* 29, no. 6 (2018): 699–702.

——. "The Pleasure of Art as a Matter of Fact." *Proceedings of the Royal Society B* 285, no. 1875 (2018): 1–3.

Nalbantian, Suzanne. "Neuroaesthetics: Neuroscientific Theory and Illustration from the Arts." *Interdisciplinary Science Reviews* 33 (2008): 357–68.

Pearce, Marcus T., Dahlia W. Zaidel, Oshin Vartanian, Martin Skov, Helmut Leder, Anjan Chatterjee, and Marcos Nadal. "Neuroaesthetics: The Cognitive Neuroscience of Aesthetic Experience." *Perspectives Psychological Science* 11 (2016): 265–79.

Poldrack, Russell A. "Can Cognitive Processes Be Inferred from Neuroimaging Data?" *Trends in Cognitive Sciences* 10, no. 2 (2006): 59–63.

Ramachandran, Vilayanur S. "The Artful Brain." In *A Brief Tour of Human Consciousness*, 40–59. New York: Pearson, [2003] 2004.

——. "The Perception of Depth from Shading." *Scientific American* 269 (1988): 76–83.

Rentschler, Ingo, Barbara Herzberger, and David Epstein, eds. *Beauty and the Brain*. Basel: Birkhäuser, 1988.

Seeley, William P. "What Is the Cognitive Neuroscience of Art . . . and Why Should We Care?" *American Society for Aesthetics* 31, no. 2 (2011): 1–4.

Segev, Idan, Luis M. Martinez, and Robert J. Zatorre. "Brain and Art." *Frontiers in Human Neuroscience* 8 (2014). https://www.frontiersin.org/articles/10.3389/fnhum.2014.00465.

Shimamura, Arthur P. *Experiencing Art: In the Brain of the Beholder*. Oxford: Oxford University Press, 2013.

Skov, Martin, and Marco Nadal. "A Farewell to Art: Aesthetics as a Topic in Psychology and Neuroscience." *Perspectives on Psychological Science* 15, no. 3 (2020): 630–42.

Skov, Martin, and Oshin Vartanian. "Introduction: What Is Neuroaesthetics?" In *Neuroaesthetics*, 1–7.

——, eds. *Neuroaesthetics*. Amityville, NY: Baywood, 2009.

——. "What Have We Learned Since *Neuroaesthetics*." Preface to the second edition of the Korean translation of *Neuroaesthetics*. N.p.: Korea Books Medio, 2019.

Slaby, Jan. "Brain Images and Neurosubjectivities." In *Critical Neuroscience: Summer School in Social and Cultural Psychiatry*. Montreal: McGill University, 2013.

Stafford, Barbara Maria. "Neuronale Ästhetik: Auf dem Weg zu einer kognitiven Bildgeschichte." In *Iconic Turn: Die neue Macht der Bilder*, ed. Christa Maar and Hubert Burda, 103–25. Cologne: DuMont, 2004.

——. "Romantic Systematics and the Genealogy of Thought: The Formal Roots of a Cognitive History of Images." *Configurations* 12 (2004): 316–44.

Starr, G. Gabrielle. *Feeling Beauty: The Neuroscience of Aesthetic Experience*. Cambridge, MA: MIT Press, 2013.

Tallis, Raymond. "The Limitations of a Neurological Approach to Art." *Lancet* 372 (2008): 19–20.

Tinio, Pablo P. L. "From Artistic Creation to Aesthetic Reception: The Mirror Model of Art." *Psychology of Aesthetics, Creativity, and the Arts* 7 (2013): 265–75.

Trentini, Bruno. "Philosophical Aesthetics and Neuroaesthetics: A Common Future?" In Kapoula and Vernet, *Aesthetics and Neuroscience*, 301–11.

Turner, Mark, ed. *The Artful Mind: Cognitive Science and the Riddle of Human Creativity*. Oxford: Oxford University Press, 2006.

Vartanian, Oshin. "Neuroimaging Studies of Making Aesthetic Products." In *Art, Aesthetics and the Brain*, ed. Joseph P. Huston, Marcos Nadal, Francisco Mora, Luigi F. Agnati, and Camilo José Cela-Conde, 174–85. Oxford: Oxford University Press, 2015.

Vassiliou, Fotini. "Aesthetic Disinterestedness in Neuroaesthetics: A Phenomenological Critique." *Aesthetic Investigations* 4, no. 1 (2020): 77–95.

Vernet, Marine. "How Art and Neuroscience Fell for Each Other." In Kapoula and Vernet, *Aesthetics and Neuroscience*, 81–89.

Vidal, Fernando. "La neuroesthétique, un esthétisme scientiste." *Revue d'Histoire des Sciences Humaines* 25, no. 2 (2011): 239–64.

Viskontas, Indre V., and Suzee E. Lee. "The Creation of Art in the Setting of Dementia." In Huston, Nadal, Mora, Agnati, and Cela-Conde, *Art, Aesthetics and the Brain*, 357–22.

Young, Kay. *Imagining Minds: The Neuro-Aesthetics of Austen, Eliot, and Hardy*. Columbus: Ohio State University Press, 2010.

Zaidel, Dahlia W. "Creativity, Brain, and Art: Biological and Neurological Considerations." *Frontiers in Human Neuroscience* 8 (2014). https://doi.org/10.3389/fnhum.2014.00389.

——. "Hemispheric Specialization, Art, and Aesthetics." In Huston, Nadal, Mora, Agnati, and Cela-Conde, *Art, Aesthetics and the Brain*, 373–82.

——. "Neuroesthetics Is Not Just About Art." *Frontiers in Human Neuroscience* 9 (2015). https://doi.org/10.3389/fnhum.2015.00080.

——. *Neuropsychology of Art: Neurological, Cognitive and Evolutionary Perspectives*. Hove, UK: Psychology Press, 2005.

Zeki, Semir. "Neural Concept Formation and Art: Dante, Michelangelo, Wagner." *Journal of Consciousness Studies* 9, no. 3 (2002): 53–76.

——. "Notes Towards a (Neurobiological) Definition of Beauty." *Gestalt Theory* 41, no. 2 (2019): 107–12.

——. *Splendors and Miseries of the Brain: Love, Creativity, and the Quest for Human Happiness*. New York: Wiley-Blackwell, 2009.

——. *A Vision of the Brain*. Oxford: Wiley-Blackwell, 1993.

Zwijnenberg, Robert. "Brains, Art and the Humanities." In Ortega and Vidal, *Neurocultures*, 293–310.

## VISUAL CULTURE

Amedi, Amir, Lofti B. Merabet, Noa Tal, and Alvaro Pascual-Leone. "Pictorial Art Beyond Sight: Revealing the Mind of a Blind Painter." In Bacci and Melcher, *Art and the Senses*, 465–79.

Beaulieu, Anne. "Images Are Not the (Only) Truth: Brain Mapping, Visual Knowledge, and Iconoclasm." *Science, Technology, & Human Values* 27, no. 1 (2002): 53–86.

Blanke, Olaf. "I and Me: Self-Portraiture in Brain Damage." In Bogousslavsky and Hennerici, *Neurological Disorders in Famous Artists, Part 2*, 22:14–29. Basel: Karger, 2007.

Bogousslavsky, Julien. "Creativity in Painting and Style in Brain-Damaged Artists." In Rose, *The Neurobiology of Painting*, 135–46.

Brincker, Maria. "The Aesthetic Stance: On the Conditions and Consequences of Becoming a Beholder." In Scarinzi, *Aesthetics and the Embodied Mind*, 117–38.

Cappa, Stefano F. "Arte e immagine corporea: considerazioni neurologiche." In Lucignani and Pinotti, *Immagini della mente*, 121–34.

Chatterjee, Anjan. "The Neuropsychology of Visual Art." In Huston, Nadal, Mora, Agnati, and Cela-Conde, *Art, Aesthetics and the Brain*, 341–56.

——. "The Neuropsychology of Visual Art: Conferring Capacity." In Rose, *The Neurobiology of Painting*, 39–49.

——. "The Neuropsychology of Visual Artistic Production." *Neuropsychologia* 42 (2004): 1568–83.

——. "Visual Art." In *Neurobiology of Sensation and Reward*, ed. Jay A. Gottfried, 391–404. New York: Taylor and Francis, 2011.

Damasio, Antonio R. "Body Images." In *Images: A Reader*, ed. Sunil Manghani, Arthur Piper, and Jon Simons, 199–202. London: Sage, 2006.

——. "Euthanasia Coaster." In *Design and Violence*, ed. Paola Antonelli and Jamer Hunt, 194–97. New York: Museum of Modern Art, 2015. Exhibition catalogue. Accessed July 21, 2020. https://www.moma.org/interactives/exhibitions/2013/designandviolence/euthanasia-coaster-julijonas-urbonas/.

——. "Film, Mind, and Emotion: The Brain Perspective: Movies on the Mind." In *Psychology and Film Since Sigmund Freud*. Berlin: Deutsche Kinemathek, Museum für Film und Fernsehen, 2006. Exhibition catalogue.

——. "Limbus." in *Fisher Museum Exhibition Catalogue*. Los Angeles: University of Southern California, 2018. Exhibition catalogue.

——. "L'omeostasi artistica." In "Van Gogh e i neuroni. L'arte come esperienza cognitiva," ed. Gloria Origgi. Special issue, *MicroMega*, no. 2 (2007): 185–89.

——. "On Seeing Things and the Power of Photography." In *See the Light: Photography, Perception, Cognition*, ed. Britt Salvesen. Los Angeles: Los Angeles County Museum of Art, 2013. Exhibition catalogue.

——. "The Photographs of Janet Sternburg." In *Fisher Museum Exhibition Catalogue*. Los Angeles: University of Southern California, 2018.

Damasio, Antonio R., and Hanna Damasio. "Infinis d'Asie," in *Catalog of the Exhibition of Jean-Baptiste Huynh*. Paris: Guimet National Museum of Asian Arts, 2019.

Dietrichs, Espen. "Peg Leg Frieda." In Rose, *The Neurobiology of Painting*, 293–99.

Dumit, Joseph. *Picturing Personhood: Brain Scans and Biomedical Identity*. Princeton, NJ: Princeton University Press, 2004.

Fedorova, Ksenia. "Neurointerfaces, Mental Imagery and Sensory Translation in Art and Science in the Digital Age." In *Invisibility in Visible and Material Culture*, ed. Asbjørn Grønstad and Øyvind Vågnes, 91–109. N.p.: Palgrave Macmillan, 2019.

Fimiani, Filippo. "Miroirs incarnés: Des images aux affects, et retour." *Miroirs* 1 (2011): 375–88.

Fingerhut, Joerg. "Twofoldness in Moving Images: The Philosophy and Neuroscience of Filmic Experience." *Projections* 14, no. 3 (2020). https://doi.org/10.3167/proj.2020.140302.

Fingerhut, Joerg, and Katrin Heimann. "Movies and the Mind: On Our Filmic Body." In *Embodiment, Enaction, and Culture: Investigating the Constitution of the Shared World*, ed. Thomas Fuchs, Christoph Durt, and Christian Tewes, 353–77. Cambridge, MA: MIT Press, 2017.

Freedberg, David. "Composition and Emotion." In Turner, *The Artful Mind*, 73–89.

——. "Empathy, Motion and Emotion." In *Wie sich Gefühle Ausdruck verschaffen: Emotionen in Nahsicht*, ed. Klaus Herding and Antje Krause Wahl, 17–51. Berlin: Driesen, 2007.

——. "Feelings on Faces: From Physiognomics to Neuroscience." In *Rethinking Emotion: Interiority and Exteriority in Premodern, Modern, and Contemporary Thought*, ed. Rüdiger Campe and Julia Weber, 289–323. Hawthorne, NY: De Gruyter, 2014.

Freedberg, David, and Vittorio Gallese. "Motion, Emotion and Empathy in Esthetic Experience." *Trends in Cognitive Sciences* 11, no. 5 (May 2007): 197–203.

Gallese, Vittorio. "Mirror Neurons and Art." In Bacci and Melcher, *Art and the Senses*, 455–63.

——. "The Problem of Images: A View from the Brain-Body." *Phenomenology and Mind*, no. 14 (2018): 70–79.

Gallese, Vittorio, and Michele Guerra. "Embodied Movies: Embodied Simulation and Film Studies." *Cinema* 3 (2012): 183–210.

——. *The Empathic Screen: Cinema and Neuroscience*. Trans. Frances Anderson. Oxford: Oxford University Press, 2019.

Gregory, Richard L. "Object Hypotheses in Visual Perception: David Marr or Cruella de Ville?" *Journal of Consciousness Studies* 6, no. 6–7 (1999): 54–56.

Harrison, Martin, ed. *Bacon and the Mind: Art, Neuroscience and Psychology*. London: Estate of Francis Bacon, 2019.

Ione, Amy. "Examining Semir Zeki's Neural Concept Formation and Art: Dante, Michelangelo, Wagner." *Journal of Consciousness Studies* 10, no. 2 (2003): 58–66.

——. "An Inquiry Into Paul Cézanne: The Role of the Artist in Studies of Perception and Consciousness." *Journal of Consciousness Studies* 7, no. 8–9 (2000): 57–74.

——. "Neurology, Synaesthesia, and Painting." In Rose, *The Neurobiology of Painting*, 69–78.

Jimenez, Marc. *Art et technosciences: Bioart et neuroesthétique*. Paris: Klincksieck, 2016.

Jokeit, Hennric, and Daniel Blochwitz. "Neuro-Aesthetics and the Iconography in Photography." *Psych Journal* 9 (2020): 444–57.

Jucker, Jean-Luc, and Johan De Smedt. "Cognition and Visual Art." In *The International Encyclopedia of Anthropology*, ed. Hilary Callan. N.p.: Wiley-Blackwell, 2018.

Kac, Eduardo, ed. *Biotechnology, Art and Culture*. Cambridge, MA: MIT Press, 2007.

Kandel, Eric R. *The Age of Insight: The Quest to Understand the Unconscious in Art, Mind, and Brain, from Vienna 1900 to the Present*. New York: Random House, 2012.

Kranjec, Alexander. "Conceptual Art Made Simple for Neuroaesthetics." *Frontiers in Human Neuroscience* 9 (2015). https://doi.org/10.3389/fnhum.2015.00267.

Lauring, Jon O. "Visual Art." In Lauring, *An Introduction to Neuroaesthetics*, 115–50.

Livingstone, Margaret. "Artisti, scienziati per istinto." In "Van Gogh e i neuroni: l'arte come esperienza cognitiva," ed. Gloria Origgi. Special issue, *MicroMega*, no. 2 (2007): 194–95.

——. *Vision and Art: The Biology of Seeing*. New York: Abrams, 2002.

Londero, Alain, Didier Bouccara, and Hervé Bozec. "The Vestibular System and Artistic Painting: A Theoretical Framework for the Study of Multi-Modal Interactions in Aesthetic Experience of Painting and Painting Viewing." In Kapoula and Vernet, *Aesthetics and Neuroscience*, 3–18.

Lucignani, Giovanni, and Andrea Pinotti, eds. *Immagini della mente: Neuroscienze, arte, filosofia*. Milan: Cortina, 2007.

Maffei, Lamberto. "I diversi sentieri della memoria e l'arte visive." In Lucignani and Pinotti, *Immagini della mente*, 69–81.

Maffei, Lamberto, and Adriana Fiorentini. *Arte e cervello*. Bologna: Zanichelli, 1995.

Mallgrave, Harry Francis. *Architecture and Embodiment: The Implications of the New Sciences and Humanities for Design*. London: Routledge, 2013.

Martindale, Colin. "Peak Shift, Prototypicality and Aesthetic Preference." *Journal of Consciousness Studies* 6, no. 6–7 (1999): 52–54.

Mather, George. *The Psychology of Visual Art: Eye, Brain and Art*. Cambridge: Cambridge University Press, 2013.

Mazzucchi, Anna, Giovanni Pesci, and Dario Trento. *Cervello e pittura*. Rome: Fratelli Palombi Editori, 2004.

McManus, Chris. "'The Anti-Developmental, the Anti-Narrative, the Anti-Historical': Mondrian as a Paradigmatic Artist for Empirical Aesthetics." *Behavioral and Brain Sciences* 36 (2013): 152–53.

Minissale, Gregory. "Conceptual Art: A Blind Spot for Neuroaesthetics?" *Leonardo* 45 (2012): 43–48.

Nicola, Ubaldo, and Klaus Podoll. *L'aura di Giorgio de Chirico: Arte emicranica e pittura metafisica*. Milan: Mimesis, 2003.

Onians, John. "Beauty in the Brain of the Beholder: Art, Neural Plasticity and Visual Pleasure." In *The Recovery of Beauty: Arts, Culture, Medicine*, ed. Corinne Saunders, Jane Macnaughton, and David Fuller, 237–53. London: Palgrave Macmillan, 2015.

——. "A Brief Natural History of Art." In *Compression vs Expression: Containing and Explaining the World's Art*, ed. John Onians, 235–49. Williamstown, MA: Clark Art Institute, 2000.

——. "The Biological Basis of Renaissance Aesthetics." In *Concepts of Beauty in Renaissance Art*, ed. Francis Ames-Lewis and Mary Rogers, 12–27. Aldershot, UK: Ashgate, 1998.

——. *European Art: A Neuroarthistory*. New Haven, CT: Yale University Press, 2016.

——. "Francis Bacon: A Neuroarthistory." In Harrison, *Bacon and the Mind*, 99–119.

——. "Greek Temple and Greek Brain." In *Body and Building: Essays on the Changing Relation of Body and Architecture*, ed. George Dodds and Robert Tavernor, 43–63. Cambridge, MA: MIT Press, 2002.

——. "A Natural Anthropology of Art." *International Journal of Anthropology* 18, no. 4 (2003): 259–64.

——. "Neuroarthistory: First Principles." Paper presented at the Neuroaesthetics Conference Goldsmith Conference in May 2005.

——. *Neuroarthistory: From Aristotle and Pliny to Baxandall and Zeki*. New Haven, CT: Yale University Press, 2007.

Origgi, Gloria, ed. "Van Gogh e i neuroni: l'arte come esperienza cognitiva." Special issue, *MicroMega*, no. 2 (2007): 185–95.

Palmer, Stephen E., Karen B. Schloss, and Jonathan Sammartino. "Visual Aesthetics and Human Preference." *Annual Review Psychology* 64 (2013): 77–107.

Pepperell, Robert. "Artworks as Dichotomous Objects: Implications for the Scientific Study of Aesthetic Experience." *Frontiers in Human Neuroscience* 9 (2015). https://doi.org/10.3389/fnhum.2015.00295.

——. “Connecting Art and the Brain: An Artist’s Perspective on Visual Indeterminacy.” *Frontiers in Human Neuroscience* 5 (2011). https://doi.org/10.3389/fnhum.2011.00084.

Pepperell, Robert C., and Alumit Ishai, “Indeterminate Artworks and the Human Brain.” In Huston, Nadal, Mora, Agnati, and Cela-Conde, *Art, Aesthetics and the Brain*, 143–57.

Prasad, Amit. “Making Images/Making Bodies: Visibilizing and Disciplining Through Magnetic Resonance Imaging (MRI).” *Science, Technology, & Human Values* 30 (2005): 291–316.

Racine, Eric, Ofek Bar-Ilan, and Judy Illes. “fMRI in the Public Eye.” *Nature Review Neuroscience* 6, no. 2 (2005): 159–64.

Ramachandran, Vilayanur S. “Response to Gombrich.” *Journal of Consciousness Studies* 7, no. 8–9 (2000): 17–20.

——. “Sharpening Up ‘The Science of Art’: An Interview with Anthony Freeman.” *Journal of Consciousness Studies* 8, no. 1 (2001): 9–29.

Ramachandran, Vilayanur S., and William Hirstein. “The Science of Art: A Neurological Theory of Aesthetic Experience.” *Journal of Consciousness Studies* 6, no. 6–7 (1999): 15–51.

Ramachandran, Vilayanur S., and Elizabeth Seckel. “Neurology of Visual Aesthetics: Indian Nymphs, Modern Art, and Sexy Beaks.” In Shimamura and Palmer, *Aesthetics Science*, 375–89.

Rollins, Mark. “What Monet Meant: Intention and Attention in Understanding Art.” In “Art, Mind and Cognitive Science,” ed. Jerrold Levinson and Jenefer Robinson. Special issue, *Journal of Aesthetics and Art Criticism* 62, no. 2 (Spring 2004): 175–88.

Rose, Frank Clifford. “The Deafness of Goya (1746–1828).” In Rose, *The Neurobiology of Painting*, 301–15.

——, ed. *The Neurobiology of Painting*. San Diego: Elsevier, 2006.

——. “The Neurology of Art: An Overview.” In Rose, *Neurology of the Arts*, 43–76.

——, ed. *Neurology of the Arts: Painting, Music, Literature*. London: Imperial College Press, 2004.

Roskies, Adina L. “Are Neuroimages Like Photographs of the Brain?” *Philosophy of Science* 74, no. 5 (2007): 860–72.

Rostrup, Nicolai. “Environmental Neuroaesthetics: Landscapes, Places and Architecture.” In Lauring, *An Introduction to Neuroaesthetics*, 185–210.

Schachter, Steven C. “The Visual Art of Contemporary Artists with Epilepsy.” In Rose, *The Neurobiology of Painting*, 119–31.

Seeley, William P. “Art, Meaning, and Aesthetics: The Case for a Cognitive Neuroscience of Art.” In Huston, Nadal, Mora, Agnati, and Cela-Conde, *Art, Aesthetics and the Brain*, 19–39.

——. “Art, Meaning, and Perception: A Question of Methods for a Cognitive Neuroscience of Art.” *British Journal of Aesthetics* 53, no. 4 (2013): 443–60.

——. “Naturalizing Aesthetics: Art and the Cognitive Neuroscience of Vision.” *Journal of Visual Art Practice* 5, no. 3 (2006): 195–213.

Siler, Todd. “Neuroart: Picturing the Neuroscience of Intentional Actions in Art and Science.” *Frontiers in Human Neuroscience* 9 (2015). https://doi.org/10.3389/fnhum.2015.00410.

Smith, Murray. *Film, Art, and the Third Culture: A Naturalized Aesthetics of Film*. Oxford: Oxford University Press, 2017.

Smith, Philip. “Fainting in Classical Art.” In Rose, *The Neurobiology of Painting*, 79–88.

Søndergaard, Morten, and Peter Weibel, eds. *Thorbjørn Lausten's Visual Systems: Magnet*. Roskilde, Denmark: MFSK—Museum of Contemporary Art, 2008. Exhibition catalogue.

Stafford, Barbara Maria. "Another Kind of Global Thinking." In Elkins, *Is Art History Global?*, 184–88.

——. *Body Criticism: Imaging the Unseen in Enlightenment Art and Medicine*. Cambridge, MA: MIT Press, 1991.

——. *Echo Objects: The Cognitive Work of Images*. Chicago: University of Chicago Press, 2007.

——. "From Genetic Perspective to Biohistory: The Ambiguities of Looking Down, Across, and Beyond." In *Sign of Life: Bio Art and Beyond*, ed. Eduardo Kac, 373–86. Cambridge, MA: MIT Press, 2007.

——. "Gewalt und Naturgeschichte: Über das dauerhafte Dilemma, wie etwas anschaulich zu machen sei." In *Vorträge aus dem Warburg-Haus* 1, 75–105. Berlin: Akademie Verlag, 1997.

——. "'Medusa' or the Physiognomy of the Earth: Humbert de Superville's Cosmological Aesthetics." *Journal of the Warburg and Courtauld Institute* 35 (1972): 308–38.

Stafford, Barbara Maria, and Horst Bredekamp. "One Step Beyond Hyperrealism." Interview/ Discussion of Hyper-Realism. In *Tate Etc.* 6 (January 2006). Accessed June 11, 2021. https://www.tate.org.uk/tate-etc/issue-5-autumn-2005/one-step-beyond.

Vidal, Fernando. "Desire, Indefinite Lifespan, and Transgenerational Brains in Literature and Film." *Theory & Psychology* 26, no. 5 (2016): 665–80.

Wade, Nicholas. "Perceptual Portraits." In Rose, *The Neurobiology of Painting*, 17–39.

Whiteley, Louise. "Resisting the Revelatory Scanner? Critical Engagements with fMRI in Popular Media." *Biosocieties* 7 (2012): 245–272.

Wingate, Richard, and Marius Kwint. "Imagining the Brain Cell: The Neuron in Visual Culture." *Nature Reviews Neuroscience* 7, no. 9 (September 2006): 745–52.

York, George K. "Localization in the Visual Brain." In Rose, *The Neurobiology of Painting*, 55–65.

Zeki, Semir. *Balthus ou la Quête de L'Essentiel: Entretiens avec Semir Zeki (Romans, Essais, Poesie, Documents)*. Paris: Les Belles Lettres, 1995.

——. "Clive Bell's 'Significant Form,' and the Neurobiology of Aesthetics." *Frontiers in Human Neuroscience* 7 (2013). https://doi.org/10.3389/fnhum.2013.00730.

——. *Inner Vision: An Exploration of Art and the Brain*. Oxford: Oxford University Press, 1999.

——. "The Neurology of Ambiguity." *Consciousness and Cognition* 13 (2004): 173–96.

Zeki, Semir, and Tomohiro Ishizu. "The 'Visual Shock' of Francis Bacon: An Essay in Neuroesthetics." In Harrison, *Bacon and the Mind*, 121–47.

Zeki, Semir, and Michael Lamb. "The Neurology of Kinetic Art." *Brain* 117 (1994): 607–36.

## PERFORMANCE

Blair, Rhonda. "4E Cognition for Directing: Thornton Wilder's Our Town and Caryl Churchill's Light Shining in Buckinghamshire." In Kemp and McConachie, *The Routledge Companion to Theatre, Performance and Cognitive Science*, 91–99.

——. "Acting and Science." In *The Cambridge Companion to Theatre and Science*, ed. Kirsten E. Shepherd-Barr, 162–75. Cambridge: Cambridge University Press, 2020.

——. "'Active Analysis—More Active Than You Know': Stanislavsky and Cognitive Science." In *The Routledge Companion to Stanislavsky*, 308–20. London: Routledge, 2014.

——"Notes on Empathy, Cognitive Neuroscience, and Theatre/Education." Special issue, *p-e-r-f-o-r-m-a-n-c-e* 2, no. 1–2 (Spring 2015). Accessed February 13, 2020.

——. "Theatre and Embodiment." *Theatre Symposium* 27 (2019): 11–23.

Blair, Rhonda, and Amy Cook, eds. *Theatre, Performance and Cognition: Languages, Bodies and Ecologies*. London: Methuen, 2016.

Bläsing, Bettina, Beatriz Calvo-Merino, Emily S. Cross, Corinne Jola, Juliane Honisch, and Catherine J. Stevens. "Neurocognitive Control in Dance Perception and Performance." *Acta Psychologica* 139 (2012): 300–308.

Bressan, Yannick. "The Theatrical Stage Setting: A Tool for Neuropsychology and Cognitive Neuroscience." In Kapoula and Vernet, *Aesthetics and Neuroscience*, 44–56.

Broadhurst, Susan. "Aesthetics, Neuroaesthetics and Embodiment: Theorising Performance and Technology." In *Music-Dance: Sound and Motion in Contemporary Discourse*, ed. Patrizia Veroli and Gianfranco Vinay, 235–48. New York: Routledge, 2017.

——. *Digital Practices: Aesthetic and Neuroesthetic Approaches to Performance and Technology*. Basingstoke, UK: Palgrave Macmillan, 2007.

Calvo-Merino, Beatriz, and Patrick Haggard. "Neuroaesthetics of Performing Arts." In Bacci and Melcher, *Art and the Senses*, 529–41.

Ciesielski, Thomas. "Contemporary Dance Theatre in Neurocognitive Perspective—Granhøj Dans Case." *Theatralia* 19, no. 2 (2016): 76–90.

Cook, Amy. "Emergence, Meaning and Presence: An Interdisciplinary Approach to a Disciplinary Question." In Kemp and McConachie, *The Routledge Companion to Theatre, Performance and Cognitive Science*, 225–34.

Davies, David. "Dancing Around the Issues: Prospects for an Empirically Grounded Philosophy of Dance." *Journal of Aesthetics and Art Criticism* 71 (2013): 195–202.

Falletti, Clelia, Gabriele Sofia, and Victor Jacono, eds. *Theatre and Cognitive Neuroscience*. London: Bloomsbury Mathuen Drama, 2016.

Fingerhut, Joerg. "Embodied Seeing-In, Empathy, and Expansionism." *Projections* 12, no. 2 (2018): 28–38.

Freedberg, David, and Antonio Pennisi. "The Body in the Picture: The Lesson of Phantom Limbs and the Origins of the BIID." *Reti, saperi, linguaggi* 9, no. 1 (January/June 2020): 5–50.

Gallese, Vittorio. "'Being Like Me': Self-Other Identity, Mirror Neurons and Empathy." In *Perspectives on Imitation: From Cognitive Neuroscience to Social Science*, ed. Susan Hurley and Nick Chater, 2:101–118. Cambridge, MA: MIT Press, 2005.

——. "A Bodily Take on Aesthetics: Performativity and Embodied Simulation." In *The Extended Theory of Cognitive Creativity: Interdisciplinary Approaches to Performativity*, ed. Antonio Pennisi and Alessandra Falzone, 135–49. Cham, Switzerland: Springer, 2020.

——. "Bodily Selves in Relation: Embodied Simulation as Second-Person Perspective on Intersubjectivity." *Philosophical Transactions of the Royal Society of London B* 369 (2014): http://dx.doi.org/10.1098/rstb.2013.0177.

——. "Brain, Body, Habit and the Performative Quality of Aesthetics." In *Habits: Pragmatist Approaches from Cognitive Neuroscience to Social Science*, ed. Italo Testa and Fausto Caruana, 376–94. Cambridge: Cambridge University Press, 2021.

——. "Embodied Simulation: Its Bearing on Aesthetic Experience and the Dialogue Between Neuroscience and the Humanities." *Gestalt Theory* 41, no. 2 (2019): 113–28.

——. "The Empathic Body in Experimental Aesthetics—Embodied Simulation and Art." In Lux and Weigel, *Empathy*, 181–99.

——. "Finding the Body in the Brain: From Simulation Theory to Embodied Simulation." In *Goldman and His Critics*, ed. Brian McLaughlin and Hilary K. Kornblith, 297–317. Oxford: Wiley, 2016.

——. "Il corpo teatrale: mimetismo, neuroni specchio, simulazione incarnata." *Culture teatrali* 16 (2008): 13–38.

——. "Mirroring, a Liberated Embodied Simulation and Aesthetic Experience." In *Mirror Images: Reflections in Art and Medicine*, 27–37. Thun, Switzerland: Verlag für moderne Kunst, Kunstmuseum Thun, 2017.

——. "Naturalizing Aesthetic Experience: The Role of (Liberated) Embodied Simulation." *Projections* 12, no. 2 (Winter 2018): 50–59.

——. "The 'Shared Manifold' Hypothesis: From Mirror Neurons to Empathy." *Journal of Consciousness Studies* 8, no. 5–7 (2001): 33–50.

——. "Visions of the Body: Embodied Simulation and Aesthetic Experience." *Aisthesis* 10, no. 1 (2017): 41–50.

Gallese, Vittorio, and Cinzia Di Dio. "Neuroesthetics: The Body in Esthetic Experience." In *The Encyclopedia of Human Behavior*, ed. Vilyanur S. Ramachandran, 2:687–93. Amsterdam: Elsevier, 2012.

Gallese, Vittorio, and David Freedberg. "Mirror and Canonical Neurons Are Crucial Elements in Esthetic Response." *Trends in Cognitive Sciences* 11, no. 10 (2007): 411.

Gallese, Vittorio, and Alessandro Gattara. "Embodied Simulation, Aesthetics, and Architecture: An Experimental Aesthetic Approach." In Robinson and Pallasmaa, *Mind in Architecture*, 161–80.

Hagendoorn, Ivar. "Cognitive Dance Improvisation: How Study of the Motor System Can Inspire Dance (and Vice Versa)." *Leonardo* 36, no. 3 (2003): 221–27.

——. "Some Speculative Hypotheses About the Nature and Perception of Dance and Choreography." *Journal of Consciousness Studies* 11, no. 3–4 (2004): 79–110.

——. "Towards a Neurocritique of Dance." *BalletTanz Yearbook* (2004): 1–6.

Ione, Amy. *Art and the Brain: Plasticity, Embodiment, and the Unclosed Circle*. Boston: Brill, 2016.

Jeannerod, Marc. "The Representing Brain: Neural Correlates of Motor Intention and Imagery." *Behavioral and Brain Sciences* 17 (1997): 187–245.

Kemp, Rick. *Embodied Acting: What Neuroscience Tells Us about the Actor*. London: Routledge, 2012.

Kemp, Rick, and Bruce McConachie, eds. *The Routledge Companion to Theatre, Performance and Cognitive Science*. London: Routledge, 2019.

Lippi, Daria, Corinne Jola, Victor Jacono, and Gabriele Sofia. "Steps Towards the Art of Placing Science in the Acting Practice: A Performance-Neuroscience Perspective." In Kapoula and Vernet, *Aesthetics and Neuroscience*, 141–64.

Lutterbie, John. *An Introduction to Theatre, Performance and Cognitive Science*. London: Methuen Drama, 2020.

McConachie, Bruce. *Evolution, Cognition, and Performance*. Cambridge: Cambridge University Press, 2015.

Papadopoulos, Dimitri. "The Imaginary of Plasticity: Neural Embodiment, Epigenetics and Ecomorphs." *Sociological Review* 59, no. 3 (2011): 432–56.

Philippe-Meden, Pierre, and Vanille Roche-Fogli, eds. *Spectacle vivant et neurosciences*. Montpellier, France: Deuxième Epoque, 2019.

Ramachandran, Vilayanur S. "Mirror Neurons and Imitation Learning as the Driving Force Behind 'The Great Leap Forward' in Human Evolution." *Edge*. Accessed February 11, 2020. www.edge.org/3rd_culture/ramachandran/ramachandran_p1.html.

Ramachandran, Vilayanur S., and Sandra Blakeslee. *Phantoms in the Brain*. New York: HarperCollins, 1998.

Ramachandran, Vilayanur S., and Edward M. Hubbard. "Synaesthesia—A Window Into Perception, Thought and Language." *Journal of Consciousness Studies* 8, no. 12 (2001): 3–34.

Scarinzi, Alfonsina, ed. *Aesthetics and the Embodied Mind: Beyond Art Theory and the Cartesian Mind-Body Dichotomy*. Dordrecht, Netherlands: Springer, 2015.

Sofia, Gabriele. "Figura(c)tions: Apprentissage de l'acteur et sciences cognitives." Ed. Mathieu Bouvier, La Manufacture, Lausanne, Switzerland, 2018. Accessed July 21, 2021. http://www.pourunatlasdesfigures.net/.

——. "The Spectator's 'Making Sense': An Interdisciplinary Research Between Theatre and Neurosciences." In *Making Sense: Merging Theory and Praxis*, ed. Bandy Lee and Lorna Collins, 175–88. Bern: Peter Lang, 2013.

——. "System Theory, Enaction and Performing Arts." In Kemp and McConachie, *The Routledge Companion to Theatre, Performance and Cognitive Science*, 195–202.

——. "Le théâtre un stimulant cérébral." *Cerveau & Psycho* 27 (March 2018): 78–82.

——. "Towards a Twentieth Century History of Relationships Between Theatre and Neuroscience." *Brazilian Journal on Presence Studies* 4, no. 2 (May/August 2014): 313–32.

## BOTTOM-UP AND EXPERIMENT-BASED NEUROSCIENTIFIC SOURCES

### AESTHETICS

Bhattacharya, Joydeep, and Hellmuth Petsche. "Drawing on Mind's Canvas: Differences in Cortical Integration Patterns Between Artists and Non-Artists." *Human Brain Mapping* 26 (2005): 1–14.

Bohrn, Isabel C., Ulrike Altmann, Oliver Lubrich, Winfried Menninghaus, and Arthur M. Jacobs. "When We Like What We Know—A Parametric fMRI Analysis of Beauty and Familiarity." *Brain and Language* 124, no. 1 (2013): 1–8.

Brown, Steven, Xiaoqing Gao, Loren Tisdelle, Simon B. Eickhoff, and Mario Liotti. "Naturalizing Aesthetics: Brain Areas for Aesthetic Appraisal Across Sensory Modalities." *NeuroImage* 58 (2011): 250–58.

Budrys, Valmantas, Kari Skullerud, Donatas Petroska, Jurate Lengveniene, and Gintaras Kaubrys. "Dementia and Art: Neuronal Intermediate Filament Inclusion Disease and Dissolution of Artistic Creativity." *European Neurology* 57 (2007): 137–44.

Cattaneo, Zaira, Carlotta Lega, Albert Flexas, Marcos Nadal, Enric Munar, and Camilo J. Cela-Conde. "The World Can Look Better: Enhancing Beauty Experience with Brain Stimulation." *Social Cognitive and Affective Neuroscience* (2013). https://doi.org/10.1093/scan/nst165.

Cela-Conde, Camilo J., Luigi Agnati, Joseph P. Huston, Francisco Mora, and Marcos Nadal. "The Neural Foundations of Aesthetic Appreciation." *Progress in Neurobiology* 94, no. 1 (2011): 39–48.

Cela-Conde, Camilo J., Juan García-Prieto, José J. Ramasco, Claudio R. Mirasso, Ricardo Bajo, Enric Munar, Albert Flexas, et al. "Dynamics of Brain Networks in the Aesthetic Appreciation." *Proceedings of the National Academy of Sciences, USA* 110 (2013): 10454–61.

Cirignotta, Fabio, C. V. Todesco, and Elio Lugaresi. "Temporal Lobe Epilepsy with Ecstatic Seizures (So-Called Dostoevsky Epilepsy)." *Epilepsia* 21 (1980): 705–10.

Coccagna, Maddalena, Pietro Avanzini, Mariagrazia Portera, Giovanni Vecchiato, Maddelena Fabbri Destro, Alessandro Vittorio Sironi, Fabrizio Salvi, et al. "Neuroaesthetics of Art Vision: An Experimental Approach to the Sense of Beauty." *Journal of Clinical Trials* 10, no. 2 (2020). https://doi.org/10.35248/2167-0870.20.10.404.

Goguen, Joseph, and Erik Myin, eds. "Art and the Brain." Special issue, *Journal of Consciousness Studies* 6, no. 6–7 (1999).

——, eds. "Art and the Brain, Part II." Special issue, *Journal of Consciousness Studies* 7, no. 8–9 (2000).

——, eds. "Art and the Brain, Part III." Special issue, *Journal of Consciousness Studies* 11, no. 3–4 (2004).

Graf, Laura K. M., and Jan R. Landwehr. "A Dual-Process Perspective on Fluency-Based Aesthetics: The Pleasure-Interest Model of Aesthetic Liking." *Personality and Social Psychology Review* 19 (2015): 395–410.

Graham, Daniel J., and Ming Meng. "Artistic Representations: Clues to Efficient Coding in Human Vision." *Visual Neurosciences* 28 (2011): 371–79.

Höfel, Lea, and Thomas Jacobsen. "Electrophysiological Indices of Processing Aesthetics: Spontaneous or Intentional Processes?" *International Journal of Psychophysiology* 65 (2007): 20–31.

Horikawa, Tomoyasu, and Yukiyasu Kamitani. "Generic Decoding of Seen and Imagined Objects Using Hierarchical Visual Features." *Nature Communications* 8 (2017). https://doi.org/10.1038/ncomms15037.

Ishizu, Tomohiro, and Semir Zeki. "The Brain's Specialized Systems for Aesthetic and Perceptual Judgment." *European Journal of Neuroscience* 37 (2013): 1413–20.

——. "A Neurobiological Enquiry Into the Origins of Our Experience of the Sublime and Beautiful." *Frontiers in Human Neuroscience* 8 (2014). https://doi.org/10.3389/fnhum.2014.00891.

Jacobs, Richard H. A. H., Remco Renken, and Frans W. Corneilissen. "Neural Correlates of Visual Aesthetics—Beauty as the Coalescence of Stimulus and Internal State." *PLOS One* 7 (2012): https://doi.org/10.1371/journal.pone.0031248.

Jacobsen, Thomas, and Lea Höfel. "Descriptive and Evaluative Judgment Processes: Behavioral and Electrophysiological Indices of Processing Symmetry and Aesthetics." *Cognitive, Affective, & Behavioral Neuroscience* 3 (2003): 289–99.

Jacobsen, Thomas, Ricarda I. Schubotz, Lea Höfel, and D. Yves von Cramon. "Brain Correlates of Aesthetic Judgment of Beauty." *NeuroImage* 29 (2006): 276–85.

Kierulf, Halfdan. "The Aetiology of Dostoevsky's Epilepsy." In Rose, *Neurology of the Arts*, 351–56.

Kirk, Ulrich. "The Modularity of Aesthetic Processing and Perception in the Human Brain: Functional Neuroimaging Studies of Neuroaesthetics." In Shimamura and Palmer, *Aesthetics Science*, 318–36.

——. "The Neural Basis of Object-Context Relationships on Aesthetic Judgment." *PLOS One* 3, no. 11 (2008): https://doi.org/10.1371/journal.pone.0003754.

Kirk, Ulrich, Martin Skov, Mark S. Christensen, and Niels Nygaard. "Brain Correlates of Aesthetic Expertise: A Parametric fMRI Study." *Brain and Cognition* 69 (2009): 306–15.

Kirk, Ulrich, Martin Skov, Oliver Hulme, Mark S. Christensen, and Semir Zeki. "Modulation of Aesthetic Value by Semantic Context: An fMRI Study." *NeuroImage* 44 (2009): 1125–32.

Kontson, Kimberly L., Murad Megjhani, Justin A. Brantley, Jesus Cruz-Garza, Sho Nakagome, Dario Robleto, Michelle White, et al. "Your Brain on Art: Emergent Cortical Dynamics during Aesthetic Experiences." *Frontiers in Human Neuroscience* 9 (2015). https://doi.org/10.3389/fnhum.2015.00626.

Leder, Helmut, Gernot Gerger, David Brieber, and Norbert Schwarz. "What Makes an Art Expert? Emotion and Evaluation in Art Appreciation." *Cognition & Emotion* 28, no. 6 (2014): 1137–47.

Locher, Paul J., Elizabeth A. Krupinski, and Alexandra Shaefer. "Art and Authenticity: Behavioral and Eye-Movement Analysis." *Psychology of Aesthetics, Creativity, and the Arts* 9 (2015): 356–67.

Meltzoff, Andrew N., and M. Keith Moore. "Explaining Facial Imitation: A Theoretical Model." *Early Development and Parenting* 6 (1997): 179–92.

Menninghaus, Winfried, Valentin Wagner, Julian Hanich, Eugen Wassiliwizky, Thomas Jacobsen, and Stefan Koelsch. "The Distancing-Embracing Model of the Enjoyment of Negative Emotions in Art Reception." *Behavioral and Brain Sciences* 20 (2017): 1–58.

Munar, Enric, Marcos Nadal, Nazareth P. Castellanos, Abert Flexas, Fernando Maestú, Claudio Mirasso, and Camilo J. Cela-Conde. "Aesthetic Appreciation: Event-Related Field and Time-Frequency Analysis." *Frontiers in Human Neuroscience* 5 (2012). https://doi.org/10.3389/fnhum.2011.00185.

Munar, Enric, Marcos Nadal, Jaume Rosselló, Albert Flexas, S. Moratti, Fernando Maestú, Gisèle Marty, and Camilo J. Cela-Conde. "Lateral Orbitofrontal Cortex Involvement in Initial Negative Aesthetic Impression Formation." *PLOS One* 7, no. 6 (2012): https://doi.org/10.1371/journal.pone.0038152.

Nadal, Marco. "The Experience of Art: Insights from Neuroimaging." *Progress in Brain Research* 204 (2013): 135–58.

Newman, George E., and Paul Bloom. "Art and Authenticity: The Importance of Originals in Judgments of Value." *Journal of Experimental Psychology: General* 14 (2012): 558–69.

Scott, Isabel M., Andrew P. Clark, Steven C. Josephson, Adam H. Boyette, Innes C. Cuthill, Ruby L. Fried, Mhairi A. Gibson, et al. "Human Preferences for Sexually Dimorphic Faces May Be Evolutionary Novel." *PNAS* 111, no. 40 (2014): 14388–93.

Skov, Martin. "Aesthetic Appreciation: The View from Neuroimaging." *Empirical Studies of the Arts* 37 (2019): 220–48.

Thakral, Preston P., Lauren R. Moo, and Scott D. Slotnick. "A Neural Mechanism for Aesthetic Experience." *NeuroReport* 23 (2012): 310–13.

Vessel, Edward A., Jonathan Stahl, Natalia Maurer, Alexander Denker, and G. Gabrielle Starr. "Personalized Visual Aesthetics." *Proceedings Vol. 9014: Human Vision and Electronic Imaging* 19 (2014): 1–8.

Vessel, Edward A., G. Gabrielle Starr, and Nava Rubin. "Art Reaches Within: Aesthetic Experience, the Self and the Default Mode Network." *Frontiers in Human Neuroscience* 7 (2013). https://doi.org/10.3389/fnins.2013.00258.

Zeki, Semir, and Jonathan Stutters. "A Brain-Derived Metric for Preferred Kinetic Stimuli." *Open Biology* 2, no. 2 (2012). https://doi.org/10.1098/rsob.120001.

## VISUAL CULTURE

Ardizzi, Martina, Francesca Ferroni, Francesca Siri, Maria A. Umiltà, Anna Cotti, Marta Calbi, Elisabetta Fadda, et al. "Beholders' Sensorimotor Engagement Enhances Aesthetic Rating of Pictorial Facial Expressions of Pain." *Psychological Research* (2018): 1–10.

Aviv, Vered. "What Does the Brain Tell Us About Abstract Art?" *Frontiers in Human Neuroscience* 8 (2014). https://doi.org/10.3389/fnhum.2014.00085.

Babiloni, Fabio, Patrizia Cherubino, Ilenia Graziani, Arianna Trettel, Francesco Infarinato, Daniela Picconi, Gianluca Borghini, et al. "Neuroelectric Brain Imaging During a Real Visit of a Fine Arts Gallery: A Neuroaesthetics Study of Seventeenth-Century Dutch Painters." *Conference Proceedings—IEEE Engineering in Medicine and Biology Society* (2013): 6179–82.

Babiloni, Francesca, Dario Rossi, Patrizia Cherubino, Arianna Trettel, Daniela Picconi, Anton Giulio Maglione, Giovanni Vecchiato, and Fabio Babiloni. "A Neuroaesthetic Study of the Cerebral Perception and Appreciation of Paintings by Titian Using EEG and Eyetracker Measurements." In *Symbiotic Interaction*, ed. Benjamin Blankertz, Giulio Jacucci, Luciano Gamberini, Anna Spagnolli, and Jonathan Freeman, 21–32. Cham, Switzerland: Springer, 2015. https://doi.org/10.1007/978-3-319-24917-9_3.

Batt, Roger, Massimiliano Palmiero, Chie Nakatani, and Cees van Leeuwen. "Style and Spectral Power: Processing of Abstract and Representational Art in Artists and Non-Artists." *Perception* 39, no. 12 (2010): 1659–71.

Battaglia, Fortunato, Sarah H. Lisanby, and David Freedberg. "Corticomotor Excitability During Observation and Imagination of a Work of Art." *Frontiers in Human Neuroscience* 5 (2011). https://doi.org/10.3389/fnhum.2011.00079.

Behrens, Marion, Pascal Nicklas, and Christian Kell. "Functional Imaging of Cognitive Processes Underlying the Perception of Contemporary Visual Art." *Conference Abstract: XII International Conference on Cognitive Neuroscience (ICON-XII)*, 2015. https://doi.org/10.3389/conf.fnhum.2015.217.00015.

Belke, Benno, Helmut Leder, and Dorothee Augustin. "Mastering Style: Effects of Explicit Style-Related Information, Art Knowledge and Affective State on Appreciation of Abstract Paintings." *Psychology Science* 48 (2006): 115–35.

Berti, Anna, and Alessia Folegatti. "Negligenza spaziale. Immagini, distorsioni e modulazioni delle rappresentazioni spaziali nella produzione grafica e artistica di pazienti cerebrolesi." In Lucignani and Pinotti, *Immagini della mente*, 207–29.

Bianucci, Raffaella, F. Marías Franco, and Otto Appenzeller. "Historical Evidence Supports El Greco's Depiction of a Neurological Condition in His Attributed Self-Portrait." *Journal of the Neurological Sciences* 372 (2017): 316–17.

Blanke, Olaf. "Visuospatial Neglect in Lovis Corinth's Self-Portraits." In Rose, *The Neurobiology of Painting*, 193–214.

Blanke, Olaf, and Isabella Maria Pasqualini. "The Riddle of Style Changes in the Visual Arts After Interference with the Right Brain." *Frontiers in Human Neuroscience* 5 (2011). https://doi.org/10.3389/fnhum.2011.00154.

Boccia, Maddelena, Sonia Barbetti, Laura Piccardi, Cecilia Guariglia, Fabio Ferlazzo, Anna Maria Giannini, and Dahlia W. Zaidel. "Where Does Brain Neural Activation in Aesthetic Responses to Visual Art Occur? Meta-Analytic Evidence from Neuroimaging Studies." *Neuroscience & Biobehavioral Reviews* 60 (2016): 65–71.

Boccia, Maddalena, Federico Nemmi, Emanuela Tizzani, Cecilia Guariglia, Fabio Ferlazzo, Gaspare Galati, and Anna Maria Giannini. "Do You Like Arcimboldo's? Esthetic Appreciation Modulates Brain Activity in Solving Perceptual Ambiguity." *Behavioural Brain Research* 278 (2015): 147–54.

Bogousslavsky, Julien. "Artistic Creativity, Style and Brain Disorders." *European Neurology* 54 (2005): 103–11.

Bogousslavsky, Julien, and François Boller, eds. *Neurological Disorders in Famous Artists, Part 1*. Basel: Karger, 2005.

Bogousslavsky, Julien, and Michael G. Hennerici, eds. *Neurological Disorders in Famous Artists, Part 2*. Basel: Karger, 2007.

Bogousslavsky, Julien, Michael G. Hennerici, Hansjörg Bäzner, and Claudio Bassetti, eds. *Neurological Disorders in Famous Artists, Part 3*. Basel: Karger, 2010.

Burton, Harold. "Visual Cortex Activity in Early and Late Blind People." *Journal of Neuroscience* 23, no. 10 (May 2003): 4005–11.

Caharel, Stéphanie, Arnaud Leleu, Christian Bernard, Maria-Pia Viggiano, Robert Lalonde, and Mohamed Rebaï. "Early Holistic Face-Like Processing of Arcimboldo Paintings in the Right Occipito-Temporal Cortex: Evidence from the N170 ERP Component." *International Journal of Psychophysiology* 90, no. 2 (2013): 157–64.

Calbi, Marta, Hava Aldouby, Ori Gersht, Nunzio Langiulli, Vittorio Gallese, and Maria A. Umiltà. "Haptic Aesthetics and Bodily Properties of Ori Gersht's Digital Art: A Behavioral and Eye-Tracking Study." *Frontiers in Psychology* 10 (2019). https://doi.org/10.3389/fpsyg.2019.02520.

Casini, Silvia. "Magnetic Resonance Imaging (MRI) as Mirror and Portrait: MRI Configurations Between Science and the Arts." *Configurations* 19, no. 1 (2011): 73–99.

Cavanagh, Patrick. "Do Artists See Their Retinas?" *Frontiers in Human Neuroscience* 5 (2011). https://doi.org/10.3389/fnhum.2011.00171.

Cela-Conde, Camilo J., Gisèle Marty, Fernando Maestú, Tomás Ortiz, Enric Munar, Alberto Fernández, et al. "Activation of the Prefrontal Cortex in the Human Visual Aesthetic Perception." *Proceedings of the National Academy of Sciences* 101 (2004): 6321–25.

Cela-Conde, Camilo J., Gisèle Marty, Enric Munar, Marcos Nadal, and Lucrecia Burges. "The 'Style Scheme' Ground Perception of Paintings." *Perceptual and Motor Skills* 95, no. 1 (2002): 91–100.

Chahboun, Sobh, Andrea Flumini, Carmen Pérez González, I. Chris McManus, and Julio Santiago. "Reading and Writing Direction Effects on the Aesthetic Appreciation of Photographs." *Laterality: Asymmetries of Body, Brain and Cognition* 22, no. 3 (2017): 313–39.

Chatterjee, Anjan. "Prospect for a Neuropsychology of Art." In Skov and Vartanian, *Neuroaesthetics*, 131–43.

Crutch, Sebastian J., and Martin N. Rossor. "Artistic Changes in Alzheimer's Disease." In Rose, *The Neurobiology of Painting*, 147–61.

Cupchik, Gerald C., Oshin Vartanian, Adrian Crawley, and David J. Mikulis. "Viewing Artworks: Contributions of Cognitive Control and Perceptual Facilitation to Aesthetic Experience." *Brain and Cognition* 70 (2009): 84–91.

De Pisapia, Nicola, Francesca Bacci, Danielle Parrott, and David Melcher. "Brain Networks for Visual Creativity: A Functional Connectivity Study of Planning a Visual Artwork." *Scientific Report* 6 (2016). https://doi.org/10.1038/srep39185.

Di Dio, Cinzia, Martina Ardizzi, and Davide Massaro. "Human, Nature, Dynamism: The Effects of Content and Movement Perception on Brain Activations during the Aesthetic Judgment of Representational Paintings." *Frontiers in Human Neuroscience* 9 (2016). https://doi.org/10.3389/fnhum.2015.00705.

Di Dio, Cinzia, Nicola Canessa, Stefano F. Cappa, and Giacomo Rizzolatti. "Specificity of Esthetic Experience for Artworks: An fMRI Study." *Frontiers in Human Neuroscience* 5 (2011). https://doi.org/10.3389/fnhum.2011.00139.

Di Dio, Cinzia, Emiliano Macaluso, and Giacomo Rizzolatti. "The Golden Beauty: Brain Response to Classical and Renaissance Sculptures." *PLOS One* 2, no. 11 (2007): https://doi.org/10.1371/journal.pone.0001201.

Di Fiore, Emanuele, and R. Di Liberti. *L'utilizzazione della stimolazione visiva subliminare nello studio dell'arte moderna*. Palermo, Italy: STASS, 1978.

Drago, Valeria, Glen R. Finney, Paul S. Foster, Alejandra Amengual, Yong Jeong, Tomoiuki Mizuno, Gregory P. Crucian, and Kenneth M. Heilman. "Spatial-Attention and Emotional Evocation: Line Bisection Performance and Visual Art Emotional Evocation." *Brain and Cognition* 66 (2008): 140–44.

Drago, Valeria, Paul S. Foster, Daniel Trifiletti, David B. FitzGerald, Gregory P. Crucian, and Kenneth M. Heilman. "'What's Inside the Art?' The Influence of Frontotemporal Dementia in Art Production." *Neurology* 10 (2006): 1285–87.

Edwards, Betty. *Drawing on the Right Side of the Brain*. 4th ed. New York: Tarcher/Penguin, [1979] 2012. Kindle.

Else, Jane E., Jason Ellis, and Elizabeth Orme. "Art Expertise Modulates the Emotional Response to Modern Art, Especially Abstract: An ERP Investigation." *Frontiers in Human Neuroscience* 9 (2015). https://doi.org/10.3389/fnhum.2015.00525.

Espinel, Carlos Hugo. "De Kooning's Late Colours and Forms: Dementia, Creativity and the Healing Power of Art." *Lancet* 347 (1996): 1096–98.

——. "Masaccio's Cripple: A Neurological Syndrome: Its Art, Medicine and Values." *Lancet* 346 (1995): 1984–86.

Fairhall, Scott L., and Alumit Ishai. "Neural Correlates of Object Indeterminacy in Art Compositions." *Consciousness and Cognition* 17 (2008): 923–32.

Freimuth, Marilyn, and Seymour Wapner. "The Influence of Lateral Organization in the Evaluation of Paintings." *British Journal of Psychology* 70 (1979): 211–18.

Gerger, Gernot, and Helmut Leder. "Titles Change the Esthetic Appreciation of Paintings." *Frontiers in Human Neuroscience* 9 (2015). https://doi.org/10.3389/fnhum.2015.00464.

Gibbons, Michelle G. "Seeing the Mind in the Matter: Functional Brain Imaging as Framed Visual Argument." *Argumentation and Advocacy* 43, no. 3–4 (2007): 175–88.

Graham, Daniel J., and Ming Meng. "Altered Spatial Frequency Content in Paintings by Artists with Schizophrenia." *i-Perception* 2 (2011): 1–9.

Graham, Daniel J., Simone Stockinger, and Helmut Leder. "An Island of Stability: Art Images and Natural Scenes—But Not Natural Faces—Show Consistent Esthetic Response in Alzheimer's-Related Dementia." *Frontiers in Psychology* 4 (2013). https://doi.org/10.3389/fpsyg.2013.00107.

Hansen, Peter, Michael Brammer, and Gemma Calvert. "Visual Preference for Art Images Discriminated with FMRI." *NeuroImage* 11, no. 5 (2000): https://doi.org/10.1016/S1053-8119(00)91669-0.

Heimann, Katrin, Sebo Uithol, Maria Alessandra Umiltà, Michele Guerra, Joerg Fingerhut, and Vittorio Gallese. "Embodying the Camera: An EEG Study on the Effect of Camera Movements on Film Spectators' Sensorimotor Cortex Activation." *PLOS One* 14, no. 3 (2019): 1–18.

Hermens, Frouke, and Johannes Zanker. "Looking at Op Art: Gaze Stability and Motions Illusions." *i-Perception* 3, no. 5 (2012): 282–304.

Herrera-Arcos, Guillermo, Jesús Tamez-Duque, Elsa Y. Acosta-De-Anda, Kevin Kwan-Loo, Mayra de-Alba, Ulises Tamez-Duque, Jose L. Contreras-Vidal, et al. "Modulation of Neural Activity During Guided Viewing of Visual Art." *Frontiers in Human Neuroscience* 11 (2017). https://doi.org/10.3389/fnhum.2017.00581.

Hu, Chuan-Peng, Yi Huang, Simon B. Eickhoff, Kaiping Peng, and Jie Sui. "Seeking the 'Beauty Center' in the Brain: A Meta-Analysis of fMRI Studies of Beautiful Human Faces and Visual Art." *bioRxiv* (2019). https://doi.org/10.1101/081539.

Huang, Mengfei, Holly Bridge, Martin J. Kemp, and Andrew J. Parker. "Human Cortical Activity Evoked by the Assignment of Authenticity When Viewing Works of Art." *Frontiers in Human Neuroscience* 5 (2011). https://doi.org/10.3389/fnhum.2011.00134.

Ishai, Alumit, Scott L. Fairhall, and Robert Pepperell. "Perception, Memory and Aesthetics of Indeterminate Art." *Brain Research Bulletin* 73 (2007): 319–24.

Johnson, Morgan G., Jeffrey A. Muday, and James A. Schirillo. "When Viewing Variations in Paintings by Mondrian, Aesthetic Preferences Correlate with Pupil Size." *Psychology of Aesthetics, Creativity, and the Arts* 4, no. 3 (2010): 161–67.

Khaw, Mel W., and David Freedberg. "Continuous Aesthetic Judgment of Image Sequences." *Acta Psychologica* 188 (2018): 213–19.

Kim, Chai-Youn, and Randolph Blake. "Brain Activity Accompanying Perception of Implied Motion in Abstract Paintings." *Spatial Vision* 20 (2007): 545–60.

Kirk, Ulrich, and David A. Freedberg. "Contextual Bias and Insulation Against Bias During Aesthetic Rating: The Roles of VMPFC and DLPFC in Neural Valuation." *Art, Aesthetics and the Brain* (2015): 158–73.

Latto, Richard, Dennis Brain, and Brendan Kelly. "An Oblique Effect in Aesthetics: Homage to Mondrian (1872–1944)." *Perception* 29 (2000): 981–87.

Latto, Richard, and Kirsty Russell-Duff. "An Oblique Effect in the Selection of Line Orientations by Twentieth Century Painters." *Empirical Studies of the Arts* 20 (2002): 49–60.

Leder, Helmut, Siegrun Bär, and Sascha Topolinski. "Covert Painting Simulations Influence Aesthetic Appreciation of Artworks." *Psychological Science* 23 (2012): 1479–81.

Leder, Helmut, Claus-Christian Carbon, and Ai-Leen Ripsas. "Entitling Art: Influence of Title Information on Understanding and Appreciation of Paintings." *Acta Psychologica* 121 (2006): 76–198.

Liu, Anli, Kelly Werner, Subhojit Roy, John Q. Trojanowski, Ursula Morgan-Kane, Bruce L. Miller, and Katherine P. Rankin. "A Case Study of an Emerging Visual Artist with Frontotemporal Lobar Degeneration and Amyotrophic Lateral Sclerosis." *Neurocase* 15 (2009): 235–47.

Locher, Paul J., Elizabeth A. Krupinski, Claudia Mello-Thoms, and Calvin F. Nodine. "Visual Interest in Pictorial Art during an Aesthetic Experience." *Spatial Vision* 21, no. 1–2 (2007): 55–77.

Massaro, Davide, Federica Savazzi, Cinzia Di Dio, David Freedberg, Vittorio Gallese, Gabriella Gilli, and Antonella Marchetti. "When Art Moves the Eyes: A Behavioral and Eye-Tracking Study." *PLOS One* 7 (2012). https://doi.org/10.1371/journal.pone.0037285.

Mastandrea, Stefano, and Maria A. Umiltà. "Futurist Art: Motion and Aesthetics as a Function of Title." *Frontiers in Human Neuroscience* 10 (2016). https://doi.org/10.3389/fnhum.2016.00201.

McKay, Sally. "No Neuron Is an Island: A Neuroaesthetic Inquiry Into Omer Fast's Mimetic Interactions." In Scarinzi, *Aesthetics and the Embodied Mind*, 315–30.

McLaughlin, John P., Pamela Dean, and Philip Stanley. "Aesthetic Preference in Dextrals and Sinistrals." *Neuropsychologia* 21 (1983): 147–53.

McLean, Christine, Stephen C. Want, and Benjamin J. Dyson. "The Role of Similarity, Sound and Awareness in the Appreciation of Visual Artwork Via Motor Simulation." *Cognition* 137 (2015): 174–81.

Melcher, David, and Francesca Bacci. "The Visual System as a Constraint on the Survival and Success of Specific Artworks." *Spatial Vision* 21, no. 3–5 (2008): 347–62.

Miall, R. Chris, and John Tchalenko. "A Painter's Eye Movements: A Study of Eye and Hand Movement During Portrait Drawing." *Leonardo* 34, no. 1 (2001): 35–40.

Miller, Bruce L., and Craig E. Hou. "Portraits of Artists: Emergence of Visual Creativity in Dementia." *Archives of Neurology* 61 (2004): 842–44.

Mizokami, Yoshinori, Takeshi Terao, Koji Hatano, Nobuhiko Hoaki, Kentaro Kohno, Yasuo Araki, Kensuke Kodama, et al. "Difference in Brain Activations During Appreciating Paintings and Photographic Analogs." *Frontiers in Human Neuroscience* 8 (2014). https://doi.org/10.3389/fnhum.2014.00478.

Moutoussis, Konstantinos, and Semir Zeki. "A Direct Demonstration of Perceptual Asynchrony in Vision." *Proceedings of the Royal Society B* 264 (1997): 393–99.

Pelowski, Matthew, Michael Forster, Pablo P. L. Tinio, Maria Scholl, and Helmut Leder. "Beyond the Lab: An Examination of Key Factors Influencing Interaction with 'Real' and 'Museum Based' Art." *Psychology of Aesthetics, Creativity, and the Arts* 11 (2017): 245–64.

Pelowski, Matthew, Patrick S. Markey, Michael Forster, Gernot Gerger, and Helmut Leder. "Move Me, Astonish Me . . . Delight My Eyes and Brain: The Vienna Integrated Model of Top-Down and Bottom-Up Processes in Art Perception (VIMAP) and Corresponding Affective, Evaluative, and Neurophysiological Correlates." *Physics of Life Reviews* 21 (2017): 80–125.

Pelowski, Matthew, Patrick S. Markey, Jon O. Lauring, and Helmut Leder. "Visualizing the Impact of Art: An Update and Comparison of Current Psychological Models of Art Experience." *Frontiers in Human Neuroscience* 10 (2016). https://doi.org/10.3389/fnhum.2016.00160.

Podoll, Klaus. "Migraine Art in the Internet: A Study of 450 Contemporary Artists." In Rose, *The Neurobiology of Painting*, 89–107.

Proverbio, Alice M., Federica Riva, and Alberto Zani. "Observation of Static Pictures of Dynamic Actions Enhances the Activity of Movement-Related Brain Areas." *PLOS One* 4, no. 5 (2009): https://doi.org/10.1371/journal.pone.0005389.

Ramachandran, Manooj, and Jeffrey K. Aronson. "The Diagnosis of Art: Arthrogryposis and Ribera's *The Clubfoot*." *Journal of the Royal Society of Medicine* 99, no. 6 (2006): 321–22.

Redies, Christoph, Jens Hasenstein, and Joachim Denzler. "Fractal-like Image Statistics in Visual Art: Similarity to Natural Scenes." *Spatial Vision* 21, 1–2 (2007): 137–48.

Root, Nicholas B., Karen Dobkins, Vilayanur S. Ramachandran, and Romke Rouw. "Echoes from the Past: Synaesthetic Colour Associations Reflect Childhood Gender Stereotypes." *Philosophical Transactions of the Royal Society of London B* 374 (2019). https://doi.org/10.1098/rstb.2018.0572.

Safran, Avinoam B., Nicolae Sanda, and José-Alain Sahel. "A Neurological Disorder Presumably Underlies Painter Francis Bacon Distorted World Depiction." *Frontiers in Human Neuroscience* 8 (2014). https://doi.org/10.3389/fnhum.2014.00581.

Savazzi, Federica, Davide Massaro, Cinzia Di Dio, Vittorio Gallese, Gabriella Gilli, and Antonella Marchetti. "Exploring Responses to Art in Adolescence: A Behavioral and Eye-Tracking Study." *PLOS One* 9, no. 7 (2014): https://doi.org/10.1371/journal.pone.0102888.

Sayim, Bilge, and Patrick Cavanagh. "What Line Drawings Reveal About the Visual Brain." *Frontiers in Human Neuroscience* 5 (2011). https://doi.org/10.3389/fnhum.2011.00118.

Sbriscia-Fioretti, Beatrice, Cristina Berchio, David Freedberg, Vittorio Gallese, and Maria A. Umiltà. "ERP Modulation During Observation of Abstract Paintings by Franz Kline." *PLOS One* 8, no. 10 (2013): https://doi.org/10.1371/journal.pone.0075241.

Schulz, Katharina, and Gregor U. Hayn-Leichsenring. "Face Attractiveness Versus Artistic Beauty in Art Portraits: A Behavioral Study." *Frontiers in Psychology* 8 (2017). https://doi.org/10.3389/fpsyg.2017.02254.

Sedgewick, Jennifer R., Bradley Weiers, Aaron Stewart, and Lorin J. Elias. "The Thinker: Opposing Directionality of Lighting Bias Within Sculptural Artwork." *Frontiers in Human Neuroscience* 9 (2015). https://doi.org/10.3389/fnhum.2015.00251.

Serrano, Cecilia, Richard Allegri, Mario Martelli, Fernando Teragano, and P. Rinalli. "Visual Art, Creativity and Dementia." *Vertex* 16 (2005): 418–29.

Silveira, Sarita, Kai Fehse, Aline Vedder, Katrin Elvers, and Kristina Hennig-Fast. "Is It the Picture or Is It the Frame? An fMRI Study on the Neurobiology of Framing Effects." *Frontiers in Human Neuroscience* 9 (2015). https://doi.org/10.3389/fnhum.2015.00528.

Silveira, Sarita, Verena Graupmann, Dieter Frey, Janusch Blautzik, Thomas Meindl, Maximilien Reiser, Cheng Chen, et al. "Matching Reality in the Arts: Self-Referential Neural Processing of Naturalistic Compared to Surrealistic Images." *Perception* 41 (2012): 569–76.

Siri, Francesca, Francesca Ferroni, Martina Ardizzi, Anna Kolesnikova, Marcella Beccaria, Barbara Rocci, Carolyn Christov-Bakargiev, et al. "Behavioral and Autonomic Responses to Real and Digital Reproductions of Works of Art." *Progress in Brain Research* 237 (2018): 201–21.

Ticini, Luca, Laura Rachman, Jerome Pelletier, and Stephanie Dubal. "Enhancing Aesthetic Appreciation by Priming Canvases with Actions That Match the Artist's Painting Style." *Frontiers in Human Neuroscience* 8 (2014). https://doi.org/10.3389/fnhum.2014.00391.

Tinio, Pablo P. L., Helmut Leder, and Marlies Strasser. "Image Quality and the Aesthetic Judgment of Photographs: Contrast, Sharpness, and Grain Teased Apart and Put Together." *Psychology of Aesthetics, Creativity, and the Arts* 5 (2010): 165–76.

Trawiński, Tobiasz, Natalie Mestry, Beth Harland, Simon P. Liversedge, Hayward J. Godwin, and Nick Donnelly. "The Spectatorship of Portraits by Naïve Beholders." *Psychology of Aesthetics, Creativity, and the Arts* 15, no. 1 (2019): 3–19.

Umiltà, Maria A., Cristina Berchio, Mariateresa Sestito, David Freedberg, and Vittorio Gallese. "Abstract Art and Cortical Motor Activation: An EEG Study." *Frontiers in Human Neuroscience* 6 (2012). https://doi.org/10.3389/fnhum.2012.00311.

Vartanian, Oshin, and Vinod Goel. "Neuroanatomical Correlates of Aesthetic Preferences for Paintings." *Neuroreport* 15 (2004): 893–97

Vartanian, Oshin, and Martin Skov. "Neural Correlates of Viewing Paintings: Evidence from a Quantitative Meta-Analysis of Functional Magnetic Resonance Imaging Data." *Brain and Cognition* 87 (2014): 52–56.

Vessel, Edward A., Natalia Maurer, Alexander H. Denker, and G. Gabrielle Starr. "Stronger Shared Taste for Natural Aesthetic Domains Than for Artifacts of Human Culture." *Cognition* 179 (October 2018): 121–31.

Vogt, Stine, and Svein Magnussen. "Expertise in Pictorial Perception: Eye-Movement Patterns and Visual Memory in Artists and Laymen." *Perception* 36, no. 1 (2007): 91–100.

Wagner, Angela, Matthias Ruf, Dieter F. Braus, and Martin H. Schmidt. "Neuronal Activity Changes and Body Image Distortions in Anorexia Nervosa." *Neuroreport* 14 (2003): 2193–97.

Walker, Francesco, Berno Bucker, Nicola C. Anderson, Daniël Schreij, and Jan Theeuwes. "Looking at Paintings in the Vincent Van Gogh Museum: Eye Movement Patterns of Children and Adults." *PLOS One* 12, no. 6 (2016): https://doi.org/10.1371/journal.pone.0178912.

Wang, Tingting, Lei Mo, Oshin Vartanian, Jonathan S. Cant, and Gerald Cupchik. "An Investigation of the Neural Substrates of Mind Wandering Induced by Viewing Traditional Chinese Landscape Paintings." *Frontiers in Human Neuroscience* 8 (2015). https://doi.org/10.3389/fnhum.2014.01018.

Wiesmann, Martin, and Alumit Ishai. "Training Facilitates Object Recognition in Cubist Paintings." *Frontiers in Human Neurosciences* 4, no. 11 (2010): 1–7.

Zaidel, Dahlia W. "Hemispheric Asymmetries in Memory for Incongruous Scenes." *Cortex* 24 (1988): 231–44.

Zeki, Semir. "Improbable Areas in Color Vision." In *The Visual Neurosciences*, ed. Leo M. Chalupa and John Simon Werner, 2:1029–39. Cambridge, MA: MIT Press, 2004.

Zeki, Semir, and Ludovica Marini. "Three Cortical Stages of Colour Processing in the Human Brain." *Brain* 121 (1998): 1669–85.

## PERFORMANCE

Alfaro, Arantxa, Ángela Bernabeu, Carlos Agulló, Jaime Parra, and Eduardo Fernánde. "Hearing Colors: An Example of Brain Plasticity." *Frontiers in Systems Neuroscience* 56, no. 9 (2015): 1–9.

Berti, Anna, and Francesca Frassinetti. "When Far Becomes Near: Re-Mapping of Space by Tool Use." *Journal of Cognitive Neuroscience* 12 (2000): 415–20.

Bottini, Gabriella, Edoardo Bisiach, Roberto Sterzi, and Giuseppe Vallar. "Feeling Touches in Someone Else's Hand." *NeuroReport* 13, no. 2 (2002): 249–52.

Brown, Steven, Michael J. Martinez, and Lawrence M. Parsons. "The Neural Basis of Human Dance." *Cerebral Cortex* 16 (2006): 1157–67.

Buccino, Giovanni, Fausta Lui, Nicola Canessa, Ilaria Patteri, Giovanna Lagravinese, Francesca Benuzzi, Carlo A. Porro, and Giacomo Rizzolatti. "Neural Circuits Involved in the Recognition of Actions Performed by Nonconspecifics: An fMRI Study." *Journal of Cognitive Neuroscience* 16 (2004): 114–26.

Calvo-Merino, Beatriz, Shantel Ehrenberg, Delia Leung, and Patrick Haggard. "Experts See It All: Configural Effects in Action Observation." *Psychological Research* 74 (2010): 447–56.

Calvo-Merino, Beatriz, Daniel E. Glaser, Julie Grèzes, Richard E. Passingham, and Patrick Haggard. "Action Observation and Acquired Motor Skills: An fMRI Study with Expert Dancers." *Cerebral Cortex* 15 (August 2005): 1243–49.

Calvo-Merino, Beatriz, Julie Grèzes, Daniel E. Glaser, Richard E. Passingham, and Patrick Haggard. "Seeing or Doing? Influence of Visual and Motor Familiarity in Action Observation." *Current Biology* 16 (2006): 1905–10.

Calvo-Merino, Beatriz, Corinne Jola, Daniel E. Glaser, and Patrick Haggard. "Towards a Sensorimotor Aesthetics of Performing Art." *Consciousness and Cognition* 17 (2008): 911–22.

Cantagallo, Anna, and Sergio Della Sala. "Preserved Insight in an Artist with Extrapersonal Spatial Neglect." *Cortex* 34 (1998): 163–89.

Christensen, Julia F., and Beatriz Calvo-Merino. "Dance as a Subject for Empirical Aesthetics." *Psychology of Aesthetics, Creativity, and the Arts* 7, no. 1 (2013): 76–88.

Christensen, Julia F., Camilo José Cela-Conde, and Antoni Gomila. "Not All About Sex: Neural and Biobehavioral Functions of Human Dance." *Annals of the New York Academy of Sciences* 1400, no. 1 (2017): 8–32.

Concerto, Carmen, Mohomad Al Sawah, Carmenrita Infortuna, David Freedberg, Eileen Chusid, Eugenio Aguglia, and Fortunato Battaglia. "Neural Circuits Underlying Motor Facilitation During Observation of Implied Motion." *Somatosensory and Motor Research* 32, no. 4 (2015): 207–10.

Concerto, Carmen, Carmenrita Infortuna, Ludovico Mineo, Manuel Pereira, David Freedberg, Eileen Chusid, Eugenio Aguglia, and Fortunato Battaglia. "Observation of Implied Motion in a Work of Art Modulates Cortical Connectivity and Plasticity." *Journal of Exercise Rehabilitation* 12, no. 5 (2016): 417–23.

Cross, Emily S., Louise Kirsch, Luca F. Ticini, and Simone Schütz-Bosbach. "The Impact of Aesthetic Evaluation and Physical Ability on Dance Perception." *Frontiers in Human Neuroscience* 5 (2011). https://doi.org/10.3389/fnhum.2011.00102.

Cruz-Garza, Jesus G., Zachary. R. Hernandez, Sargoon Nepaul, Karen K. Bradley, and Jose L. Contreras-Vidal. "Neural Decoding of Expressive Human Movement from Scalp Electroencephalography (EEG)." *Frontiers in Human Neuroscience* 8 (2014). https://doi.org/10.3389/fnhum.2014.00188.

Daparti, Elena, Marco Iosa, Patrick Haggard. "A Dance to the Music of Time: Aesthetically-Relevant Changes in Body Posture in Performing Art." *PLOS One* 4 (2009): 1–11.

Di Pellegrino, Giuseppe, Luciano Fadiga, Leonardo Fogassi, Vittorio Gallese, and Giacomo Rizzolatti. "Understanding Motor Events: A Neurophysiological Study." *Experimental Brain Research* 91 (1992): 176–80.

Ehrsson, H. Henrik, Birgitta Rosén, Anita Stockselius, Christina Ragnö, Peter Köhler, and Göran Lundborg. "Upper Limb Amputees Can Be Induced to Experience a Rubber Hand as Their Own." *Brain* 131, no. 12 (December 2008): 3443–52.

Ferrari, Pier Francesco, Vittorio Gallese, Giacomo Rizzolatti, and Leonardo Fogassi. "Mirror Neurons Responding to the Observation of Ingestive and Communicative Mouth Actions in the Monkey Ventral Premotor Cortex." *European Journal of Neuroscience* 17 (2003): 1703–14.

Fogassi, Leonardo, Pier Francesco Ferrari, Benno Gesierich, Stefano Rozzi, Fabian Chersi, and Giacomo Rizzolatti. "Parietal Lobe: From Action Organization to Intention Understanding." *Science* 308 (2005): 662–67.

Fogassi, Leonardo, and Vittorio Gallese. "The Neural Correlates of Action Understanding in Non-Human Primates." In *Mirror Neurons and the Evolution of Brain and Language*, ed. Maxim I. Stamenov and Vittorio Gallese, 13–15. Amsterdam: John Benjamins, 2002.

Frith, Christopher D., and Eraldo Paulesu. "The Physiological Basis of Synesthesia." In Simon Baron-Cohen and John E. Harrison, *Synaesthesia, Classic and Contemporary Readings*, 123–47. Oxford: Blackwell, 1997.

Gallese, Vittorio, Luciano Fadiga, Leonardo Fogassi and Giacomo Rizzolatti. "Action Recognition in the Premotor Cortex." *Brain* 199 (1996): 593–609.

Giraux, Pascal, and Angela Sirigu. "Les mains dans la tête." *La Recherche*, no. 366 (July/August 2003): 60–63.

Goller, Aviva I., Kerrie Richards, Steven Novak, and Jamie Ward. "Mirror-Touch Synaesthesia in the Phantom Limbs of Amputees." *Cortex* 49, no. 1 (2013): 243–51.

Grafton, Scott T., Michael A. Arbib, Luciano Fadiga, and Giacomo Rizzolatti. "Localization of Grasp Representations in Humans by Positron Emission Tomography. 2. Observation Compared with Imagination." *Experimental Brain Research* 112 (1996): 103–11.

Hadley, Bree J. "Disability, Disabled Dance Audiences and the Dilemma of Neuroaesthetic Approaches to Perception and Interpretation." In *Dance, Disability, and Law: Invisible Difference*, ed. Sarah Whatley, Charlotte Waelde, Shawn Harmon, Abbe Brown, Karen Wood, and Hetty Blades, 293–315. Bristol: Intellect, 2018.

Iacoboni, Marco, Istvan Molnar-Szakacs, Vittorio Gallese, Giovanni Buccino, John Mazziotta, and Giacomo Rizzolatti. "Grasping the Intentions of Others with One's Own Mirror Neuron System." *PLOS Biology* 3 (2005): 529–35.

Jola, Corinne, Angharad Davis, and Patrick Haggard. "Proprioceptive Integration and Body Representation: Insights into Dancers' Expertise." *Experimental Brain Research* 213, no. 2–3 (2011): 257–65.

Jola, Corinne, Shantel Ehrenberg, and Dee Reynolds. "The Experience of Watching Dance: Phenomenological-Neuroscience Duets." *Phenomenology and the Cognitive Sciences* 11 (2012): 17–37.

Kirsch, Louise P., Kim A. Drommelschmidt, and Emily S. Cross. "The Impact of Sensorimotor Experience on Affective Evaluation of Dance." *Frontiers in Human Neuroscience* 7 (2013). https://doi.org/10.3389/fnhum.2013.00521.

Milán, Emilio Gómez, Oscar Iborra, María José de Cordoba, Verónica Juarez-Ramos, María Angeles Rodríguez Artacho, and José Luis Rubio. "The Kiki-Bouba Effect: A Case of Personification and Ideasthesia." *Journal of Consciousness Studies* 20, no. 2 (2013): 84–102.

Mineo, Ludovico, Alexander Fetterman, Carmen Concerto, Michael Warren, Carmenrita Infortuna, David Freedberg, Eileen Chusid, Eugenio Aguglia, and Fortunato Battaglia. "Motor Facilitation During Observation of Implied Motion: Evidence for a Role of the Left Dorsolateral Prefrontal Cortex." *International Journal of Psychophysiology* 128 (2018): 47–51.

Nordin-Bates, Sanna M., Imogen J. Walker, Jo Baker, Jocelyn Garner, Cinzia Hardy, Sarah Irvine, Corinne Jola, Helen Laws, and Peta Blevins. "Injury, Imagery and Self-Esteem in Dance: Healthy Minds in Injured Bodies?" *Journal of Dance Medicine & Science* 15, no. 2 (2011): 76–85.

Orgs, Guido, Jans-Henryk Dombrowski, Martin Heil, and Petra Jansen-Osmann. "Expertise in Dance Modulates Alpha/Beta Event-Related Desynchronization During Action Observation." *European Journal of Neuroscience* 27 (2008): 3380–84.

Piechowski-Jozwiak, Bartlomiej, François Boller, and Julien Bogousslavsky. "Universal Connection Through Art: Role of Mirror Neurons in Art Production and Reception." *Behavioral Sciences* 7, no. 2 (2017). https://doi.org/10.3390/bs7020029.

Ramachandran, Vilayanur S., and David Brang. "Sensations Evoked in Patients with Amputation from Watching an Individual Whose Corresponding Intact Limb Is Being Touched." *Archives of Neurology* 66, no. 10 (2009): 1281–84.

Ramachandran, Vilanayur S., and Paul D. McGeoch. "Phantom Penises in Transsexuals: Evidence of an Innate Gender-Specific Body Image in the Brain." *Journal of Consciousness Studies* 15, no. 1 (2008): 5–16.

Ramachandran, Vilayanur S., and Diane Rogers-Ramachandran. "Synaesthesia in Phantom Limbs Induced with Mirrors." *Proceedings of the Royal Society B* 263, no. 1369 (1996): 377–86.

Rizzolatti, Giacomo, Luciano Fadiga, Massimo Matelli, Valentino Bettinardi, Eraldo Paulesu, Daniela Perani, and Ferrucio Fazio. "Localization of Grasp Representations in Humans by PET: 1. Observation Versus Execution." *Experimental Brain Research* 111 (1996): 246–52.

Seeley, William P. "Movement, Gesture and Meaning: A Sensorimotor Model for Audience Engagement with Dance." In *Moving Imagination: Explorations of Gesture and Inner Movement in the Arts*, ed. Helena de Preester, 51–68. Amsterdam: John Benjamins, 2013.

Sevdalis, Vassilis, and Peter E. Keller. "Captured by Motion: Dance, Action Understanding, and Social Cognition." *Brain and Cognition* 77 (2011): 231–36.

Shriki, Oren, Yaniv Sadeh, and Jamie Ward. "The Emergence of Synaesthesia in a Neuronal Network Model Via Changes in Perceptual Sensitivity and Plasticity." *PLOS Computational Biology* 12, no. 7 (2016): https://doi.org/10.1371/journal.pcbi.1004959.

Umiltà, Maria Alessandra, Ludovic Escola, Irakli Intskirveli, Franck Grammont, Magali Rochat, Fausto Caruana, Ahmad Jezzini, Vittorio Gallese, and Giacomo Rizzolatti. "When Pliers Become Fingers in the Monkey Motor System." *PNAS* 105, no. 6 (2008): 2209–13.

Umiltà, Maria Alessandra, Evelyne Kohler, Vittorio Gallese, Leonardo Fogassi, Luciano Fadiga, Christian Keysers, and Giacomo Rizzolatti. "'I Know What You Are Doing': A Neurophysiological Study." *Neuron* 31 (2001): 155–65.

## COMPLEMENTARY BIBLIOGRAPHY

Adam, Hans Christian, ed. *Karl Blossfeldt 1865–1932*. Cologne: Taschen, 1999.

Albano, Caterina, Ken Arnold, and Marina Wallace. *Head On: Art with the Brain in Mind*. London: Artakt, 2002.

Alexander, Richard D. "Evolution of the Human Psyche." In *The Human Revolution: Behavioural and Biological Perspectives on the Origins of Modern Humans*, ed. Paul Mellars and Chris Stringer, 455–513. Princeton, NJ: Princeton University Press, 1989.

Anglada-Tort, Manuel, and Martin Skov. "What Counts as Aesthetics in Science? A Bibliometric Analysis and Visualization of the Scientific Literature from 1970 to 2018." *Psychology of Aesthetics, Creativity, and the Arts* (2020). https://doi.org/10.1037/aca0000350.

Anzieu, Didier. *The Skin-Ego*. Trans. Naomi Segal. London: Routledge, [1995] 2018.

Appleton, Jay. *The Experience of Landscape*. New York: Wiley, 1996.

Aristotle. *Poetics*. Trans. Anthony Kenny. Oxford: Oxford University Press, 2013.

Arnheim, Rudolf. "The Art of Psychotics." *Art Psychotherapy* 4 (1977): 113–20.

Arnheim, Rudolf. *The Genesis of a Painting: Picasso's Guernica*. Berkeley: University of California Press, [1962] 1973.

——. *The Power of the Center: A Study of Composition in the Visual Arts*. Los Angeles: University of California Press, 1982.

——. *Visual Thinking*. Berkeley: University of California Press, 1969.

Arráez-Aybar, Luis-Alfonso, Pedro Navia-Álvarez, Talia Fuentes-Redondo, and José-L Bueno-López. "Thomas Willis, a Pioneer in Translational Research in Anatomy (on the 350th Anniversary of *Cerebri Anatome*)." *Journal of Anatomy* 226, no. 3 (2015): 289–300.

Artaud, Antonin. "Van Gogh, the Man Suicided by Society." [1947]. In *Antonin Artaud: Anthology*, ed. Jack Hirschman, 135–63. San Francisco: City Lights, 1965.

Bacci, Francesca, and David Melcher, eds. *Art and the Senses*. New York: Oxford University Press, 2011.

Barad, Karen. *Meeting the Universe Halfway: Quantum Physics and the Entanglement of Matter and Meaning*. Durham, NC: Duke University Press, 2007.

Barba, Eugenio. *On Directing and Dramaturgy: Burning the House*. Trans. Judy Barba. New York: Routledge, [2009] 2010.

——. *The Paper Canoe: A Guide to Theatre Anthropology*. Trans. Richard Fowler. London: Routledge, [1992] 1995.

——. *Theatre—Solitude, Craft, Revolt*. Trans. Judy Barba. Aberystwyth, UK: Back Mountain, [1996] 1999.

Barnes, Albert. *The Art in Painting*. 3rd ed. Merion, PA: Barnes Foundation, [1925] 1937.

Barrett, Louise. "The Evolution of Cognition: A 4E Perspective." In Newen, De Bruin, and Gallagher, *The Oxford Handbook of 4E Cognition*, 719–34.

Barthes, Roland. "From Work to Text." In *The Rustle of Language*, 56–64. Trans. Richard Howard. Berkeley: University of California Press, [1971] 1989.

Baudrillard, Jean. *America*. Trans. Chris Turner. London: Verso, [1986] 1988.

Baumgarten, Alexander Gottlieb. *Aesthetica*. Hamburg: F. Meiner, [1750] 2007.

Bell, Jameson Kismet. "The Performativity of a Historical Brain Event: Revisiting 1517 Strassburg." In Littlefield and Johnson, *The Neuroscientific Turn*, 49–70.

Benjamin, Walter. "The Work of Art in the Age of Mechanical Reproduction." In *Illuminations*, trans. Harry Zohn, ed. Hannah Arendt, 166–95. Boston: First Mariner, [1935] 2019.

Bennett, Max R., and Peter Michael Stephan Hacker. *Philosophical Foundation of Neuroscience*. Oxford: Blackwell, 2003.

Berenson, Bernard. *Aesthetics and History in the Visual Arts*. New York: Pantheon, 1948.

——. *Seeing and Knowing*. N.p.: Forgotten Books, [1953] 2017.

Bleeker, Maaike. *Visuality in the Theatre: The Locus of Looking*. London: Palgrave Macmillan, 2008.

Boehm, Gottfried. "Die Wiederkehr der Bilder." In *Was ist ein Bild?*, ed. Gottfried Boehm, 11–38. Munich: Fink, 1994.

Bohr, Niels. "The Atomic Theory and the Fundamental Principles Underlying the Description of Nature." In *Atomic Theory and the Fundamental Principles Underlying the Description of Nature*, 102–19. Cambridge: Cambridge University Press, [1929] 1934.

Borges, Jorge Luis. "On Exactitude in Science" [1960]. In *Collected Fictions*, trans. Andrew Hurley, 325. New York: Penguin, 1999.

Brandt, Anthony. "Form and Meaning in Art." In Turner, *The Artful Mind*, 171–88.

Bratu Hansen, Miriam. *Cinema and Experience: Siegfried Kracauer, Walter Benjamin, and Theodor W. Adorno*. Berkeley: University of California Press, 2012.

Brecht, Bertolt. "Alienation Effects in Chinese Acting" [1937]. In *Brecht on Theater: The Development of an Aesthetic*, trans. and ed. John Willett, 91–99. New York: Hill and Wang, [1957] 1964.

Breidbach, Olaf, ed. *Natur der Ästhetik—Ästhetik der Natur*. Vienna: Springer, 1997.

Brockman, John. *The Third Culture: Beyond the Scientific Revolution*. New York: Simon & Schuster, 1995.

Bryson, Norman. "Natural Attitude." In *Vision and Painting: The Logic of the Gaze*, 1–12. New Haven, CT: Yale University Press, 1983.

Bullot, Nicolas J., and Rolf Reber. "The Artful Mind Meets Art History: Toward a Psycho-Historical Framework for the Science of Art Appreciation." *Behavioral and Brain Sciences* 36 (2013): 123–80.

Bullot, Nicolas J., William P. Seeley, and Stephen Davies. "Art and Science: A Philosophical Sketch of Their Historical Complexity and Codependence." *Journal of Aesthetics and Art Criticism* 75, no. 4 (2017): 453–63.

Burke, Edmund. *A Philosophical Enquiry Into the Origin of Our Ideas of the Sublime and Beautiful*. Ed. Paul Guyer. Oxford: Oxford University Press, [1757] 1990.

Canguilhem, Georges. *The Normal and the Pathological*. Introduction by Michel Foucault. Trans. Carolyn R. Fawcett. New York: Zone Books, [1966] 1991.

Canuto, Ricciotto. "L'Art Cérébriste." *Le Figaro*, February 9, 1914, 1–2.

Cappelletto, Chiara. "L'attore, una figura chiasmatica." In *Filosofie sull'attore*, ed. Katia L. Angioletti, 63–78. Milan: LED, 2010.

——. "The Mask: Expressing the Scattered Human Identity." In *Okologien des Ausdrucks*, ed. Franz Fehrenbach and Matthew Vollgraf, 165–78. Berlin: De Gruyter, 2022.

——. "Weightless: Face, Faciality and Body." In *Journal of History and Philosophy of Art*, ed. Shen Yubing, 221–42. N.p.: Commercial Press, 2020.

Cartwright, Lisa. *Screening the Body: Tracing Medicine's Visual Culture*. Minneapolis: University of Minnesota Press, 1995.

Carvalho, John M. *Thinking with Images: An Enactivist Aesthetics*. New York: Routledge, 2019.

Casini, Silvia. *Giving Bodies Back to Data: Image Makers, Bricolage, and Reinvention in Magnetic Resonance Technology*. Cambridge, MA: MIT, 2021.

Cavanagh, Patrick. "The Artist as Neuroscientist." *Nature* 434 (March 17, 2005): 301–7.

Cassirer, Ernst. "Eidos and Eidolon: The Problem of Beauty and Art in the Dialogues of Plato." In *The Warburg Years (1919–1933): Essays on Language, Art, Myth and Technology*, trans. Steve G. Lofts and Antonio Calcagno, 214–43. New Haven, CT: Yale University Press, [1924] 2013.

——. "Nature-Concepts and Culture-Concepts." In *The Logic of the Humanities*, trans. Clarence Smith Howe, 117–58. New Haven, CT: Yale University Press, 1961.

Changeux, Jean-Pierre, Nicole Rouillé, and Jean-Claude Boyer. *Les passions de l'âme: Peintures des XVIIe et XVIIIe siècles de la collection Changeux*. Paris: Odile Jacob, 2006.

Churchland, Patricia S. *Touching a Nerve: The Self as Brain*. New York: Norton, 2013.

Clark, Andy. *Natural-Born Cyborgs: Minds, Technologies and the Future of Human Intelligence*. New York: Oxford University Press, 2003.

——. "Re-Inventing Ourselves: The Plasticity of Embodiment, Sensing, and Mind." *Journal of Medicine and Philosophy: A Forum for Bioethics and Philosophy of Medicine* 32, no. 3 (2007): 263–82.

——. *Supersizing the Mind: Embodiment, Action and Cognitive Extension*. New York: Oxford University Press, 2008.

Clynes, Mandred, and Nathan Kline. "Cyborgs and Space." *Astronautics* 5, no. 9 (September 1960): 26–27, 74–76.

Coleridge, Samuel Taylor. *Biographia Literaria*. London, 1817.

Colombetti, Giovanna. *The Feeling Body: Affective Science Meets the Enactive Mind*. Cambridge, MA: MIT Press, 2014.

Colombetti, Giovanna, and Eder Zavala. "Are Emotional States Based in the Brain? A Critique of Affective Brainocentrism from a Psychological Perspective." *Biology and Philosophy* 34, no. 45 (2019). https://doi.org/10.1007/s10539-019-9699-6.

Conrad, Waldemar. "Der ästhetische Gegenstand: Eine phänomenologische Studie." *Zeitschrift für Ästhetik und Allgemeine Kunstwissenschaft* 3 (1908): 71–118.

Courtine, Jean-Jacques, and Claudine Haroche. *Histoire du visage: exprimer et taire ses émotions (du XVI siècle au début du XIXe siècle)*. Paris: Payot et Rivages, 1988.

Cozens, Alexander. *A New Method of Assisting the Invention in Drawing Original Compositions of Landscape*. London, 1785.

Crary, Jonathan. *Techniques of the Observer: On Vision and Modernity in the Nineteenth Century*. Cambridge, MA: MIT Press, 1990.

Da Vinci, Leonardo. *Libro di pittura*. Facsimile edition of the Codex Urbinas Vaticanus Latinus 1270. Ed. Carlo Pedretti. Florence: Giunti, 1995. Translation in Claire Farago, *Leonardo Da Vinci's Paragone: A Critical Interpretation with a New Edition of the Text in the Codex Urbinas*. Leiden: Brill, 1992.

——. *A Treatise on Painting*. Trans. John Francis Rigaud. London: George Bell and Sons, [1540] 1892.

Damasio, Antonio R. *Descartes' Error: Emotion, Reason, and the Human Brain*. New York: Avon, 1994.

——. *Descartes' Error: Emotion, Reason, and the Human Brain*. Reprint ed. New York: Penguin, [1994] 2005.

——. *Looking for Spinoza: Joy, Sorrow, and the Feeling Brain*. New York: Harvest, 2003.

Damasio, Hanna, Thomas Grabowski, Randall Frank, Albert M. Galaburda, and Antonio R. Damasio. "The Return of Phineas Gage: Clues About the Brain from the Skull of a Famous Patient." *Science* 264, no. 5162 (1994): 1102–5.

Danto, Arthur C. "The Artworld." *Journal of Philosophy* 61, no. 19 (1964): 571–84.

——. "Seeing and Showing." *Journal of Aesthetics and Art Criticism* 59 (2001): 1–9.

Darwin, Charles. *The Expression of Emotion in Man and Animals*. With introduction, afterword, and commentaries by Paul Ekman. 4th ed. Oxford: Oxford University Press, [1872] 2009.

Davis, Whitney. "Neurovisuality." In "Evaluating Neuroaesthetics," ed. Todd Cronan. Special issue, *nonsite.org*, no. 2 (2011). Accessed July 21, 2020. https://nonsite.org/neurovisuality/.

De Chirico, Giorgio. "Sull'arte metafisica." In *Il meccanismo del pensiero: Critica, polemica, autobiografia, 1911–1943*, ed. Maurizio Fagiolo, 85–86. Turin: Einaudi, 1985.

De Condillac, Etienne Bonnot. *Traité des sensations*. [1754]. Paris: A. Hatier, 1949.

Decroux, Étienne. *Words on Mime*. Trans. Mark Piper. Claremont, CA: Mime Journal, [1963] 1985.

Deleuze, Gilles. *Cinema 2: The Time-Image*. Trans. Hugh Tomlinson and Robert Galeta. Minneapolis: University of Minnesota Press, [1985] 1997.

——. *Francis Bacon: The Logic of Sensation*. Trans. Daniel W. Smith. London: Continuum [1981], 2003.

Deleuze, Gilles, and Félix Guattari. *A Thousand Plateaus: Capitalism and Schizophrenia*. Trans. Brian Massumi. London: Athlone, [1980] 1988.

Deragna, Sergio. *Grafologia e neuroscienze: Testo teorico di semeiotica grafologica morettiana*. Rome: CE.DI.S., 2002.

Deregowski, Jan B. "On Seeing a Picture for the First Time." *Leonardo* 9, no. 1 (1976): 19–23.

Descola, Philippe. *Beyond Nature and Culture*. Trans. Janet Lloyd. Chicago: University of Chicago Press, [2005] 2014.

——. "La double vie des images." In *Penser l'image II: Anthropologies du visuel*, ed. Emmanuel Alloa, 131–45. Paris: Les Presses du Réel, 2015.

Dewey, John. "Affective Thought." In *Later Works*, ed. A. J. Boydstone. Vol. 2. Carbondale-Edwardsville: Southern Illinois University Press, [1926] 1988.

Di Marco, Silvia. "Inside the Body: Medical Imaging and Visual Arts." *PSICOART*, no. 2 (2011–2012). Accessed June 2021. https://psicoart.unibo.it/

Dilthey, Wilhelm. *Selected Works Volume I: Introduction to the Human Sciences*. Ed. Rudolf A. Makkreel and Frithjof Rodi. Princeton, NJ: Princeton University Press, [1883] 1989.

Dissanayake, Ellen. "The Artification Hypothesis and Its Relevance to Cognitive Science, Evolutionary Aesthetics and Neuroaesthetics." *Cognitive Semiotics* 5 (2009): 148–73.

——. *Homo Aestheticus: Where Art Comes From and Why*. New York: Free Press, 1992.

——. *What Is Art For?* Seattle: University of Washington Press, 1988.

Dostoevsky, Fyodor M. *The Idiot*. Trans. David McDuff. London: Penguin, [1869] 2004.

Du Bos, Jean-Baptiste. *Critical Reflections on Poetry, Painting and Music*. Ed. James O. Young and Margaret Cameron. London: Brill, [1719] 2021.

Dubuffet, Jean. *Brussels, Galerie Le Diable par la queue*. Paris: Desjobert, 1949. Exhibition catalogue.

Duchenne de Boulogne, Guillaume-Benjamin. *The Mechanism of Human Facial Expression*. Trans. and ed. R. Andrew Cuthbertson. Cambridge: Cambridge University Press, [1862] 1990.

D'Udine, Jean. *L'art et le geste*. Paris: F. Alcan, 1910.

——. *De la corrélation des sons et des couleurs en art*. Paris, 1897.

Dufrenne, Mikel. *L'oeil et l'oreille*. Montreal: L'Hexagone, 1987.

Dumit, Joseph. "Is It Me or My Brain? Depression and Neuroscientific Facts." *Journal of Medical Humanities* 24, no. 1–2 (2003): 35–47.

Ede, Siân. *Art and Science*. London: Tauris, 2005.

Edwards, David. *Artscience: Creativity in the Post-Google Generation*. Cambridge, MA: Harvard University Press, 2009.

Ehrenberg, Alain. "Le Cerveau 'social': Chimère épistémologique et vérité sociologique." *Esprit*, no. 1 (January 2008): 79–103.

Ekman, Paul. Afterword to *The Expression of Emotion in Man and Animals*, by Charles Darwin, 363–93. Oxford: Oxford University Press, [1872] 2009.

Elkins, James. "Art History as a Global Discipline." In Elkins, *Is Art History Global?*, 3–23.

——, ed. *Is Art History Global?* New York: Routledge, 2007.

——. *Pictures and Tears*. New York: Routledge, [2001] 2005.

——. *Six Stories from the End of Representation: Images in Painting, Photography, Astronomy, Microscopy, Particle Physics and Quantum Mechanics, 1980–2000*. Stanford, CA: Stanford University Press, 2008.

Ernst, Max. *Écritures: Avec cent vingt illustrations extraites de l'oeuvre de l'auteur*. Paris: Gallimard, 1970.

Exner, Sigmund. "Die Physiologie des Fliegens und Schwebens in den bildenden Künsten" [1882]. In Breidbach, *Natur der Ästhetik—Ästhetik der Natur*, 225–59.

Falcinelli, Riccardo. *Guardare, pensare, progettare: neuroscienze per il design*. Viterbo: Stampa Alternativa e Graffiti, 2011.

Fechner, Gustav Theodor. "Vergleichende Anatomie der Engel" [1875]. In Breidbach, *Natur der Ästhetik—Ästhetik der Natur*, 196–220.

Feyerabend, Paul K. *Wissenschaft als Kunst*. Frankfurt: Suhrkamp, 1984.

Fisher, John A. "Environmental Aesthetics." In *The Oxford Handbook of Aesthetics*, ed. Jerrold Levinson, 667–78. Oxford: Oxford University Press, 2003.

Foster, Hal, Rosalind Krauss, Yve-Alain Bois, and Benjamin H. D. Buchloh. *Art Since 1900: Modernism, Antimodernism, Postmodernism*. 2 vols. Vol. 2 reprinted with corrections. New York: Thames and Hudson, 2004.

Foucault, Michel. *The Birth of the Clinic: An Archaeology of Medical Perception*. Trans. A. M. Sheridan. N.p.: Routledge, [1963] 1973.

Frankel, Felice. *L'incanto della scienza*. Ed. Maria Perosino. Turin: Codice, 2004.

Frazzetto, Giovanni. "Neural Networking in Manhattan." *Nature* 451 (February 14, 2008). https://doi.org/10.1038/451772a.

Frazzetto, Giovanni, and Suzanne Anker. "Neuroculture." *Perspective* 10, no. 815 (2009): 815–20.

Freedberg, David. "From Absorption to Judgment: Empathy in Aesthetic Response." In Lux and Weigel, *Empathy*, 139–80.

——. "The Problem of Classicism: Ideology and Power." *Art Journal* 47, no. 1 (Spring 1988): 7–9.

——. *The Power of Images: Studies in the History and Theory of Response*. Chicago: University of Chicago Press, 1989.

Freud, Sigmund. "The Moses of Michelangelo" [1914]. In *The Standard Edition of the Complete Works of Sigmund Freud*, trans. and ed. James Strachey, 209–38. Vol. 13. London: Vintage, 2001.

Fried, Michael. *Absorption and Theatricality: Painting and Beholder in the Age of Diderot*. Chicago: University of Chicago Press, 1980.

Fry, Roger. *Reflections on British Painting*. London: Faber and Faber, 1934.

Gaius Plinius Secundus (Pliny the Elder). "Book XXXV: Painting, Sculpture and Architecture." In *Natural History: A Selection*, trans. John F. Healy. London: Penguin, 1992.

García-López, Pablo, Virginia García-Marín, and Miguel Freire. "The Histological Slides and Drawings of Cajal." *Frontiers in Neuroanatomy* 4, no. 9 (2010). https://doi.org/10.3389/neuro.05.009.2010.

Gardner, Howard, and Ellen Winner. "Children's Conceptions (and Misconceptions) of the Arts." In Howard Gardner, *Art, Mind, and Brain: A Cognitive Approach to Creativity*, 106–9. New York: Basic Books, 1982.

Garrels, Gary. "Three Toads in the Garden: Line, Color and Form." In *Willem de Kooning: The Late Paintings, the 1980s*. Minneapolis: Walker Art Center, 1995. Exhibition catalogue.

Gastaut, Henri. "La maladie de Vincent van Gogh envisagée à la lumière des conceptions nouvelles sur l'épilepsie psychomotrice." *Annales Médico-Psychologiques* 114 (1956): 196–238.

Geiger, Moritz. "Die psychische Bedeutung der Kunst." In *Zugänge zur Ästhetik*, 57–135. Leipzig: Der Neue Geist, 1928.

——. "Vom Dilettantismus im künstlerischen Erleben." In *Zugänge zur Ästhetik*, 1–42.

Gell, Alfred. *Art and Agency: An Anthropological Theory*. Oxford: Oxford University Press, 1998.

Goethe, Johann Wolfgang von. "Fortunate Encounter." [1794]. In *Scientific Studies*, trans. and ed. Douglas Miller, 18–21. New York: Suhrkamp, 1988.

——. "Observations on the Laocoön." [1798]. In *Goethe on Art*, trans. and ed. John Gage, 78–88. Berkeley: University of California Press, 1980.

Golden, Lauren. "Science, Darwin and Art History." In *Raising the Eyebrow: John Onians and World Art Studies: An Album Amicorum in His Honour*, ed. Lauren Golden, 79–90. Oxford: Archaeopress, 2001.

Gombrich, Ernst Hans. *Art and Illusion: A Study in the Psychology of Pictorial Representation*. Princeton, NJ: Princeton University Press, 1960.

——. "Concerning 'The Science of Art': Commentary on Ramachandran and Hirstein." *Journal of Consciousness Studies* 7, no. 8–9 (2000): 17.

——. "Psychoanalysis and the History of Art." In *Meditations on a Hobby Horse and Other Essays*, 30–44. London: Phaidon, 1963.

——. *The Sense of Order: A Study in the Psychology of Decorative Art*. London: Phaidon, 1979.

Gombrich, Ernst Hans, and Didier Eribon. *Looking for Answers: Conversations on Art and Science*. New York: Abrams, [1991] 1993.

Goodman, Nelson. *The Languages of Art*. Indianapolis: Hackett, 1968.

Goodnough, Robert. "Pollock Paints a Picture," *Art News* (Spring 1951). Accessed February 11, 2020. https://www.artnews.com/art-news/news/pollock-paints-a-picture-2127.

Greenberg, Clement. "Avant-Garde and Kitsch." In *Art and Culture: Critical Essays*, 1–33. Boston: Beacon, [1939] 1961.

Gregory, Richard L. "The Confounded Eye." In Gregory and Gombrich, *The Illusion in Nature and Art*, 49–96.

——. "Perceptions as Hypotheses." *Philosophical Transactions of the Royal Society of London B* 290, no. 1038 (1980): 181–97.

Gregory, Richard L., and Ernst Hans Gombrich, eds. *The Illusion in Nature and Art*. London: Duckworth, 1973.

Gruber, David R. *Brain Art and Neuroscience: Neurosensuality and Affective Realism*. New York: Routledge, 2020.

Haraway, Donna J. "A Cyborg Manifesto: Science, Technology, and Socialist-Feminism in the Late Twentieth Century" [1985]. In *Simians, Cyborgs, and Women: The Reinvention of Nature.* New York: Routledge, 1991.

Haeckel, Ernst. *Art Forms in Nature.* Munich: Prestel, [1904] 1998.

Hein, George E. "John Dewey and Albert C. Barnes: A Deep and Mutually Rewarding Friendship." *Dewey Studies* 1, no. 1 (Spring 2017): 44–78.

Herder, Johann Gottfried. *Sculpture: Some Observations on Shape and Form from Pygmalion's Creative Dream.* Trans. Jason Gaiger. Chicago: University of Chicago Press, [1778] 2002.

Hildebrand, Adolf. "The Problem of Form in the Fine Arts" [1893]. In Mallgrave and Ikonomou, *Empathy, Form, and Space*, 227–80.

Hixson, Joseph R. "New Seeing-Eye Machines . . . Look Inside Your Body, Can Save Your Life." *Vogue*, July 1983.

Hoel, Aud Sissel Aurora. "Images as Active Powers for Reality: A Simondonian Approach to Medical Imaging." In *Dynamis of the Image*, ed. Emmanuel Alloa and Chiara Cappelletto, 287–310. Berlin: De Gruyter, 2020.

Hoffmann, Roald. "Making Sense of the Image in the Nanoworld." In *Blow-up: Images from the Nanoworld, Seen by Lucia Covi.* Bologna: Damiani Editore, 2006.

Hughes, Robert. *Goya.* New York: Knopf, 2003.

Husserl, Edmund. *The Crisis of European Sciences and Transcendental Phenomenology: An Introduction to Phenomenological Philosophy.* Trans. David Carr. Evanston, IL: Northwestern University Press, [1936] 1970.

——. *Ideas Pertaining to a Pure Phenomenology and to a Phenomenological Philosophy, Volume II: Phenomenological Analyses Relating to the Problems of Constitution.* Trans. Richard Rojcewicz and André Schuwer. Dordrecht, Netherlands: Kluwer, [1912–] 1989.

Iacoboni, Marco. *I neuroni specchio: Come capiamo ciò che fanno gli altri.* Turin: Bollati Boringhieri, 2008.

Iannetti, Giandomenico, Tim V. Salomons, Massieh Moayedi, André Mouraux, and Karen D. Davis. "Beyond Metaphor: Contrasting Mechanism of Social and Physical Pain." *Trends in Cognitive Sciences* 17, no. 8 (2013): 371–78.

Ihde, Don, and Lambros Malafouris. "Homo Faber Revisited: Postphenomenology and Material Engagement Theory." *Philosophy & Technology* 32, no. 2 (2019): 195–214.

Ingold, Tim. *Making: Anthropology, Archaeology, Art and Architecture.* New York: Routledge, 2013.

Jacob, Pierre, and Marc Jeannerod. *Ways of Seeing: The Scope and Limits of Visual Cognition.* Oxford: Oxford University Press, 2003.

James, Tony. *Dream, Creativity, and Madness in Nineteenth-Century France.* Oxford: Clarendon Press, 1995.

James, William. *Principles of Psychology.* 2 vols. New York: Dover, [1890] 1950.

Jamison, Kay Redfield. *Touched with Fire: Manic-Depressive Illness and the Artistic Temperament.* New York: Simon & Schuster, 1993.

Jeannerod, Marc. *La nature de l'esprit: Science cognitive et cerveau.* Paris: Odile Jacob, 2002.

Jewanski, Jörg, Sean A. Day, and Jamie Ward. "A Colorful Albino: The First Documented Case of Synaesthesia, by Georg Tobias Ludwig Sachs in 1812." *Journal of the History of the Neurosciences* 18, no. 3 (2009): 293–303.

Kandel, Eric R. "Calcium and the Control of Synaptic Strength by Learning." *Nature* 293 (1981): 697–700.

Kandel, Eric R., James H. Schwartz, and Thomas M. Jessell. *Principles of Neural Science*, 3rd ed. Norwalk, CT: Appleton and Lange, 1991.

——. *Principles of Neural Science*, 5th ed. New York: McGraw-Hill, 2013.

Kandinsky, Wassily. *Concerning the Spiritual in Art*. Trans. Michael T. H. Sadler. New York: Dover, [1912] 1977.

——. *Point and Line to Plane*. Trans. Howard Dearstyne. New York: Dover, [1926] 1979.

Kant, Immanuel. *Critique of Judgment*. Ed. Nicholas Walker. Trans. James Creed Meredith. Rev. ed. Oxford: Oxford University Press, [1790] 2009.

——. *Critique of Pure Reason*. Trans. and ed. Paul S. Guyer and Allen W. Wood. Cambridge: Cambridge University Press, [1781] 1998.

Katz, David. *War Greco astigmatisch? Eine psychologische Studie zur Kunstwissenschaft*. Berlin, Boston: De Gruyter, [1914] 2021.

Kemp, Martin. "From Science in Art to the Art of Science." *Nature* 434 (March 17, 2005): 308–9.

Kierkegaard, Søren. *Papers and Journals: A Selection*. Trans. Alastair Hannay. New York: Penguin, [1846] 1996.

Klee, Felix, ed. *Paul Klee: His Life and Work in Documents, Selected from Posthumous Writings and Unpublished Letters*. Trans. Richard and Clara Winston. New York: Braziller, [1960] 1962.

Klee, Paul. *The Diaries of Paul Klee 1898–1918*. Ed. Felix Klee. Berkeley: University of California Press, [1957] 1968.

Konorski, Jerzy. *Conditioned Reflexes and Neuron Organization*. New York: Cambridge University Press, 1948.

Krois, John Michel. "Enactivism and Embodiment in Picture Acts." In *Sehen und Handeln*, ed. Horst Bredekamp and John Michel Krois, 3–19. Berlin: Akademie Verlag, 2011.

Lanier, Jaron. "What Information Is Given by a Veil?" *Journal of Consciousness Studies* 6, no. 6–7 (1999): 65–68.

Latour, Bruno. "What Is Iconoclash? Or Is There a World Beyond the Image Wars?" In *Beyond the Image Wars in Science, Religion and Art*, ed. Peter Weibel and Bruno Latour, 14–37. Cambridge, MA: MIT Press, 2002.

Lennox, William G. *Epilepsy and Related Disorders*. With Margaret A. Lennox. 2 vols. London: J. and A. Churchill, 1960.

Leslie, Charles Robert. *Memoirs of the Life of John Constable Composed Chiefly of His Letters*. Ed. Jonathan Mayne. London: Phaidon, 1951.

Lessing, Gotthold Ephraim. *Laocoön: An Essay on the Limits of Painting and Poetry*. Trans. Edward Allen McCormick. Baltimore: Johns Hopkins University Press, [1766] 1984.

Lindner, Gustav Adolf. *Lehrbuch der empirischen Psychologie als induktiver Wissenschaft*. 3rd ed. Vienna, 1872.

Linnik, Irina Vladimirovna. "Un tableau de Jacques de Bellange nouvellement découvert." *Revue de l'Art* 20 (1973): 65–70.

Lipps, Theodor. *Ästhetik: Psychologie des Schönen und der Kunst*. 2 vols. Hamburg: L. Voss, 1903–1906.

Locke, John. *An Essay Concerning Human Understanding*. Ed. Pauline Phemister. Oxford: Oxford University Press, [1690] 2008.

Long, Rose-Carol Washton, ed. "Franz Marc, letter to Reinhard Piper, April 20, 1910, and to Wassily Kandinsky, October 5, 1912." In *German Expressionism: Documents from the End of the Wilhelmine Empire to the Rise of National Socialism*, with the assistance of Ida Katherine Rigby and contributions by Stephanie Barron, Rosemarie Haag Bletter, and Peter Chametzky, trans. and ed. Nancy Roth, 48–49. Berkeley: University of California Press, 1995.

Lux, Vanessa, and Sigrid Weigel, eds. *Empathy: Epistemic Problems and Cultural-Historical Perspectives of a Cross-Disciplinary Concept*. London: Palgrave MacMillan, 2017.

Malabou, Catherine. *Ontology of the Accident: An Essay on Destructive Plasticity*. Trans. Carolyn Shread. Cambridge: Polity, [2009] 2012.

——. *Plasticité*. Paris: Léo Scheer, 2000.

——. *What Should We Do with Our Brain?* Trans. Sebastian Rand. New York: Fordham University Press, [2004] 2008.

Mallgrave, Harry Francis, and Eleftherios Ikonomou, eds. *Empathy, Form, and Space: Problems in German Aesthetics, 1873–1893*. Trans. Harry Francis Mallgrave and Eleftherios Ikonomou. Los Angeles: Getty Center for the History of Art and the Humanities, 1993.

Malraux, André. "The Museum Without Walls." In *The Voices of Silence*, trans. Stuart Gilbert, 13–130. Princeton, NJ: Princeton University Press, 1978.

Manzoni, Tullio. *Aristotele e il cervello: le teorie del più grande biologo dell'antichità nella storia del pensiero scientifico*. Rome: Carocci, 2007.

Marey, Étienne-Jules. *La méthode graphique dans les sciences expérimentales et principalement en physiologie et en médecine*. Paris: G. Masson, 1878.

Maturana, Humberto R., and Francisco J. Varela. *Autopoiesis and Cognition: The Realization of the Living*. Dordrecht. Netherlands: D. Reidel, 1980.

McManus, Chris. "Measuring the Culture of C. P. Snow's *Two Cultures*." *Empirical Studies of the Arts* 24, no. 2 (2006): 219–27.

Melchior-Bonnet, Sabine. *The Mirror: A History*. Trans. Katharine H. Jewett. New York: Routledge, [1995] 2009.

Merleau-Ponty, Maurice. "Eye and Mind." [1964] In *The Primacy of Perception*, ed. James M. Edie, trans. Carleton Dallery, 159–90. Evanston, IL: Northwestern University Press, 1964.

——. *Phenomenology of Perception*. Trans. Colin Smith. London: Routledge, [1945] 2002.

Merritt, Richard K. "From Memory Arts to the New Code Paradigm: The Artist as Engineer of Virtual Information Space and Virtual Experience." *Leonardo* 34, no. 5 (2001): 403–8.

Meulders, Michael. *Helmholtz: From Enlightenment to Neuroscience*. Trans. Laurence Garey. Cambridge, MA: MIT Press, [2001] 2012.

Mitchell, W. J. T. "Realism and the Digital Image." In *Travels in Intermedia: ReBlurring the Boundaries*, ed. Bernd Herzogenrath, 46–62. Hanover, NH: Dartmouth College Press, 2012.

——. "Showing Seeing: A Critique of Visual Culture." In *Art History, Aesthetics, Visual Studies*, ed. Michael Ann Holly and Keith Moxey, 231–50. Williamstown, MA: Clark Art Institute, 2002.

Mitchell, W. J. T., and Marquard Smith. "Mixing It Up: The Media, the Senses, and Global Politics; Interview with W. J. T. Mitchell." In *Visual Culture Studies: Interviews with Key Thinkers*, ed. Marquard Smith, 33–48. London: Sage, 2008.

Moore, Gregory. "Art and Evolution: Nietzsche's Physiological Aesthetics." *British Journal for the History of Philosophy* 10, no. 1 (2002): 109–26.

Moore, Henry. "The Sculptor Speaks." In *Henry Moore on Sculpture*, 193–98. London: Macdonald, 1966.

Morelli, Giovanni. *The Borghese and Doria-Pamfili Galleries in Rome*. Trans. Constance Joyce Ffoulkes. London: J. Murray, [1897] 1988.

Morini, Simona, and Maria Perosino, eds. *Specchi: Scienza e coscienza allo specchio*. Turin: Codice, 2006. Exhibition catalogue.

Morris, Desmond. *The Biology of Art: A Study of the Picture-Making Behavior of the Great Apes and Its Relationship to Human Art*. London: Methuen, 1969.

Morrison, India. "Mirror Neurons and Cultural Transmission." In *Mirror Neurons and the Evolution of Brain and Language*, ed. Maxim I. Stamenov and Vittorio Gallese, 333–40. Amsterdam: John Benjamins, 2002.

Mouraux, André, and Gian Domenico Iannetti. "The Search for Pain Biomarkers in the Human Brain." *Brain* 141 (2018): 3290–307.

Naccache, Lionel. *Le nouvel inconscient: Freud, Christophe Colomb des neurosciences*. Paris: Odile Jacob, 2006.

Neidich, Warren. *The Glossary of Cognitive Activism (for a Not So Distant Future)*. Berlin: Archive, 2019.

Newen, Albert, Leon De Bruin, and Shaun Gallagher, eds. *The Oxford Handbook of 4E Cognition*. Oxford: Oxford University Press, 2018.

Nietzsche, Friedrich. *The Dawn of Day*. Trans. Johanna Volz. London: T. Fisher Unwin, [1879–81] 1903.

Nijholt, Anton. *Brain Art: Brain-Computer Interfaces for Artistic Expression*. Cham, Switzerland: Springer, 2019.

Nissel, Jenny, Angelina Hawley-Dolan, and Ellen Winner. "Can Young Children Distinguish Abstract Expressionist Art from Superficially Similar Art by Children and Animals?" *Journal of Cognition and Development* 17, no. 1 (January 2016): 18–29.

Noë, Alva. *Varieties of Presence*. Cambridge, MA: Harvard University Press, 2012.

O'Connell, Garret, Janet De Wilde, Jane Haley, Kirsten Shuler, Burkhard Schafer, Peter Sandercock, and Joanna M. Wardlaw. "The Brain, the Science and the Media." *EMBO Reports* 12, no. 7 (2011): 630–36.

O'Connor, Cliodhna, and Helene Joffe. "How Has Neuroscience Affected Lay Understandings of Personhood? A Review of the Evidence." *Public Understanding of Science* 22, no. 3 (2013): 254–68.

Ortega, Francesco, and Fernando Vidal. *Being Brains: Making the Cerebral Subject*. New York: Fordham University Press, 2017.

——, eds. *Neurocultures: Glimpses Into an Expanding Universe*. Frankfurt: Peter Lang, 2011.

Pallasmaa, Juhani. "Body, Mind, And Imagination: The Mental Essence of Architecture." In Robinson and Pallasmaa, *Mind in Architecture*, 51–74.

——. *The Eyes of the Skin: Architecture and the Senses*. Chichester, UK: Wiley, [1996] 2012.

Palmier, Jean-Michel, François Aubral, Anthony Rowley, Pierre Vallaud, and Jean-Nöel von der Weid. *L'art dégénéré: Une exposition sous le IIIe Reich*. Paris: Jacques Bertoin, [1992] 2002.

Palomino de Castro y Velasco, Antonio. "Don Diego Velázquez de Silva." In *El Museo pictórico y escala óptica*. 3 vols. Madrid: Aguilar, [1796] 1988.

Panofsky, Erwin. "On the Problem of Describing and Interpreting Works of the Visual Arts." [1932] *Critical Inquiry* 38, no. 3 (Spring 2012): 467–82.

——. *Galileo as a Critic of the Arts*. The Hague: Springer, [1954] 2014.

Petit, Jean-Luc. "La spazialità originaria del corpo proprio. Fenomenologia e neuroscienze." In *Neurofenomenologia: Le scienze della mente e la sfida dell'esperienza cosciente*, ed. Massimiliano Cappuccio, 163–94. Milan: Bruno Mondadori, 2006.

Pickersgill, Martyn, Sarah Cunningham-Burley, and Paul Martin. "Constituting Neurologic Subjects: Neuroscience, Subjectivity and the Mundane Significance of the Brain." *Subjectivity* 4, no. 3 (2011): 346–65.

Philostratus. *The Life of Apollonius of Tyana: Volume I, Books 1–5*. Trans. F. C. Conybeare. Cambridge, MA: Harvard University Press, [2000] 2012.

Pinker, Steven. *How the Mind Works*. New York: Norton, 1997.

Pinotti, Andrea. "Du Bos, l'empatia e i neuroni-specchio." In *Jean-Baptiste Du Bos e l'estetica dello spettatore*, ed. Luigi Russo, 203–12. Palermo, Italy: Aesthetica Preprint, 2005.

——. "Motricità. Empatia, corpo e movimento dalla Kunstwissenschaft alla neuroestetica." In *Logiche della corporeità*, ed. Raffaele Bruno. Milan: Franco Angeli, 2008.

——. "Neuroestetica, estetica psicologica, estetica fenomenologica: le ragioni di un dialogo." *Rivista di Estetica* 37, no. 1 (2008): 147–68.

Pitts-Taylor, Victoria. *The Brain's Body: Neuroscience and Corporeal Politics*. Durham, NC: Duke University Press, 2016.

Plessner, Helmuth. *Levels of Organic Life and the Human: An Introduction to Philosophical Anthropology*. Trans. Millay Hyatt. New York: Fordham University Press, [1928] 2019.

——. "Zur Anthropologie des Schauspielers." [1948] In *Gesammelte Schriften*, ed. Günter Dux, Odo Marquard, and Elisabeth Ströker, 399–418. Frankfurt: Suhrkamp, 1982.

Poussin, Nicolas. *Collection de lettres de Nicolas Poussin*. Paris, [1647] 1824.

Preciado, Paul B. *Counter-Sexual Manifesto*. Trans. Kevin Gerry Dunn. New York: Columbia University Press, [2000] 2018.

Preti, Giulio. *Retorica e logica*. Turin: Einaudi, 1968.

Prince, Susan H., ed. *Antisthenes of Athens: Texts, Translations, and Commentary*. Ann Arbor: University of Michigan Press, 2015.

Printz, Bellamy, Jaron Lanier, and Vilayanur S. Ramachandran. *Truth, Technology and the Visual/Virtual World*. Cleveland: Cleveland Public Library, 2006.

Prinzhorn, Hans. *The Artistry of the Mentally Ill: A Contribution to the Psychology and Psychopathology of Configuration*. Trans. Eric von Brockdorff. New York: Springer, [1922] 1972.

Rampley, Matthew. "Fish, Volcanoes and the Art of Brains: Review of John Onians European Art: A Neuroarthistory." *Journal of Art Historiography* 15 (2016). Accessed March 3, 2021. https://arthistoriography.files.wordpress.com/2016/10/rampley-review-2.pdf.

——. "From Big Art Challenge to a Spiritual Vision: What 'Global Art History' Might Really Mean." In Elkins, *Is Art History Global?*, 188–202.

Réja, Marcel. *L'Art chez les fous*. Paris: Société du Mercure de France, 1907.

Reicke, Rudolf, ed. *Lose Blätter aus Kants Nachlass*. Königsberg, Germany: Beyer, 1895.

Rekand, Tiina. "Absinthe: The Nervous System and Painting." In Rose, *The Neurobiology of Painting*, 271–78.

Riegl, Alois. *Late Roman Art Industry*. Trans. Rolf Winkes. Rome: Giorgio Bretschneider, [1901] 1985.

Ritchin, Fred. *After Photography*. New York: Norton, 2009.

Rizzolatti, Giacomo, and Corrado Sinigaglia. *Mirrors in the Brain: How Our Minds Share Actions, Emotions, and Experience*. Trans. Frances Anderson. Oxford: Oxford University Press, [2006] 2008.

Robinson, Derek, and Klaus Podoll. *Migraine Art: The Migraine Experience from Within*. Berkeley, CA: North Atlantic, 2009.

Robinson, Sarah, and Juhani Pallasmaa, eds. *Mind in Architecture: Neuroscience, Embodiment, and the Future of Design*. Cambridge, MA: MIT Press, 2015. Kindle.

Rorschach, Hermann. *Psychodiagnostics: A Diagnostic Test Based on Perception*. Trans. Paul Lemkau. 2nd ed. New York: Grune & Stratton, [1921] 1942.

Rose, Nikolas, and Joelle M. Abi-Rached. *Neuro: The New Brain Sciences and the Management of the Mind*. Princeton, NJ: Princeton University Press, 2013.

Sacks, Oliver. "The Landscape of His Dreams." In *An Anthropologist on Mars: Seven Paradoxical Tales*. New York: First Vintage, [1995] 1996.

Sandblom, Philip. *Creativity and Disease: How Illness Affects Literature, Art and Music*. 12th ed. New York: Marion Boyars, 1995.

Sandrone, Stefano, Marco Bacigaluppi, Marco R. Galloni, Stefano F. Cappa, Andrea Moro, Marco Catani, Massimo Filippi, et al. "Weighing Brain Activity with the Balance: Angelo Mosso's Original Manuscripts Come to Light." *Brain* 137, no. 2 (February 2014): 621–33.

Satz, Aura. "'The Conviction of Its Existence': Silas Weir Mitchell, Phantom Limbs and Phantom Bodies in Neurology and Spiritualism." In *Neurology and Modernity*, ed. Laura Salisbury and Andrew Shail, 113–29. Basingstoke, UK: Palgrave Macmillan, 2010.

Schaeffer, Jean-Marie. *Adieu à l'esthétique*. Paris: PUF, 2000.

Schmarsow, August. "Gemeinschaft der Sinnesgebiete im schöpferischen Akt." *Zeitschrift für Ästhetik und allgemeine Kunstwissenschaft* 24 (1930): 1–15.

Sebeok, Thomas. "Prefigurements of Art." In *Readings in Zoosemiotics*, ed. Timo Maran, Dario Martinelli, and Aleksei Turovski, 194–246. Berlin: De Gruyter, 2011.

Selfe, Lorna. *Nadia: A Case of Extraordinary Drawing Ability in an Autistic Child*. Cambridge, MA: Academic Press, 1977.

Shapiro, Roberta, and Nathalie Heinich. "When Is Artification?" *Contemporary Aesthetics* 4 (2012). Accessed July 20, 2020. https://digitalcommons.risd.edu/liberalarts_contempaesthetics/volo/iss4/9.

Shaw, Joyce Cutler. "'The Anatomy Lesson': The Body, Technology and Empathy." *Leonardo* 27, no. 1 (1994): 29–38.

Shimamura, Arthur P. "Muybridge in Motion: Travels in Art, Psychology and Neurology." *History of Photography* 26 (2002): 341–50.

Shimamura, Arthur P., and Stephen E. Palmer, eds. *Aesthetics Science: Connecting Minds, Brains, and Experience*. New York: Oxford University Press, 2012.

Siebers, Tobin. *Disability Aesthetics*. Ann Arbor: University of Michigan Press, 2010.

Simondon, Gilbert. *On the Mode of Existence of Technical Objects*. Trans. **Cecile Malaspina and John Rogove**. Minneapolis: University of Minnesota Press, [1958] 2017.

Solso, Robert L. *The Psychology of Art and the Evolution of the Conscious Brain*. Cambridge, MA: MIT Press, 2003.

Soussloff, Catherine M. "Jackson Pollock's Post-Ritual Performance: Memories Arrested in Space." *Drama Review* 48, no. 1 (Spring 2004): 60–78.

Stein, Gertrude. *Picasso*. London: Dover, [1938] 1984.

Sulloway, Frank J. *Freud, Biologist of the Mind: Beyond the Psychoanalytic Legend*. Cambridge, MA: Harvard University Press, [1979] 1982.

Switzer, Ellen. "Brain Work." *Vogue* November 1983.

Tallis, Raymond. *Aping Mankind: Neuromania, Darwinitis and the Misrepresentation of Humanity*. Durham, NC: Acumen, 2011.

Thornton, Davi Johnson. *Brain Culture: Neuroscience and Popular Media*. New Brunswick, NJ: Rutgers University Press, 2011.

Tinbergen, Nikolaas. *Social Behaviour in Animals with Special Reference to Vertebrates*. New York: Wiley, 1953.

Vasari, Giorgio. "A Description of the Works of Titian of Cadore, Painter." In *The Lives of the Artists*, trans. Julia Conaway Bondanella and Peter Bondanella, 489–508. New York: Oxford University Press, [1550] 1991.

Verstegen, Ian. "Revisiting Arnheim and Gombrich in Social Scientific Perspective." *Journal of Aesthetics and Art Criticism* 76, no. 1 (Winter 2018): 45–55.

Vico, Giambattista. *The New Science of Giambattista Vico: Unabridged Translation of the Third Edition, with the Addition of "Practic of the New Science"*. Trans. Thomas Goddard Bergin and Max Harold Fisch. Ithaca, NY: Cornell University Press, [1774] 2015.

Vidal, Fernando, "Brainhood, Anthropological Figure of Modernity." *History of the Human Sciences* 22, no. 1 (2009): 5–36.

Vischer, Robert. "On the Optical Sense of Form: A Contribution to Aesthetics" [1873]. In Mallgrave and Ikonomou, *Empathy, Form, and Space*, 89–124.

Voland, Eckart, and Karl Grammer, eds. *Evolutionary Aesthetics*. Berlin: Springer, 2003.

Volkelt, Johannes. *System der Ästhetik*. Vol. 1. Munich: Beck, 1905.

Warburg, Aby. "Francesco Sassetti's Last Injunction to His Sons." In *The Renewal of Pagan Antiquity: Contributions to the Cultural History of the European Renaissance*, trans. David Britt, 223–62. Los Angeles: Getty Research Institute for the History of Art and the Humanities, [1907] 1999.

Wickens, Andrew P. *A History of the Brain: From Stone Age Surgery to Modern Neuroscience*. New York: Psychology Press, 2015.

Wittgenstein, Ludwig. "Lectures on Aesthetics." [1938] In *Lectures and Conversations: On Aesthetics, Psychology and Religious Belief*, ed. Cyril Barrett, 1–40. Berkeley: University of California Press, 1966.

Wolfe, Charles T., ed. *Brain Theory: Essays in Critical Neurophilosophy*. New York: Palgrave Macmillan, 2013.

Wölfflin, Heinrich. "How One Should Photograph Sculpture" [1896]. Trans. Geraldine A. Johnson. *Art History* 36, no. 1 (2013): 52–71.

——. "Prolegomena to a Psychology of Architecture" [1886]. In Mallgrave and Ikonomou, *Empathy, Form, and Space*, 149–92.

Wood, Christopher S. "E. H. Gombrich's 'Art and Illusion: A Study in the Psychology of Pictorial Representation' " [1960]. *Burlington Magazine* 151, no. 1281 (December 2009): 836–39.

Worringer, Wilhelm. *Abstraction and Empathy: A Contribution to the Psychology of Style*. Trans. Michael Bullock. Chicago: Ivan R. Dee, [1907] 1997.

# APPENDIX

## ARTWORKS ON THE BRAIN

Helen Chadwick, *Self-Portrait* (1991), from the series *Meat Lamps*, installation, photographic transparency, glass, aluminum frame, and electric lights, 20 × 17.6 × 4.7 in. Artist holding what is implied to be her brain in her hands.

Joyce Cutler Shaw, *Memory Picture No. 12* (1992), from *The Anatomy Lesson*, photograph. Coronal section MRI superimposed over oval of the artist's face.

Karen Norberg, *Knitted Brain* (1994), colored cotton yarn, 9 in. Anatomically precise brain.

Susan Aldworth, *Cogito Ergo Sum 3* (2002). Twenty fMRI scans of the artist's brain. Scale altered and scans doctored to include words and images tied to her everyday experience.

Marta de Menezes, *Functional Portraits* (2002). fMRI images of the artist's brain activity while she draws.

Andrew Carnie and Richard Wingate, *Magic Forest* (2002), time-based/sculptural work. Visitors walked through a forest of "floating trees"—that is, neurons captured in the act of transmitting signals to dendrites.

Darij Kreuh and Janez Janša, *Brainscore* (2003), neurogame/inter-media, through frames, as a performance by the artists. Blend of virtual reality technology and brain signal processing whereby two operators, acting through avatars, could communicate by brain waves and eye movements.

Marc Didou, *Eco* (2004), bronze sculpture, 86.6 × 64.6 × 32.2 in. Two specular heads.

Mariko Mori, *Wave UFO* (2005), large architectural sculpture. Fusion of real-time computer graphics, brainwave technology, sound, and architectural engineering.

Constance Jacobson, *Brains* (2008), watercolor and ink. Reenacts the work of Ramón y Cajal.

Marjorie Taylor, *Warm Glow* or *Fabric MRI Bill's Brain* (2009), rug from wool fabric strips, 5 × 6 ft. Twenty-three horizontal sections from an fMRI of a human brain, reproduced roughly but unmistakably.

Jan Fabre, *Pietas* (2011). Statue of the Madonna holding Christ's body. White marble reenactment of Michelangelo's *Pietà*.

Dario Ghibaudo, *I peccati capitali sono otto* (The deadly sins are eight) (2011), in the fifteenth room of his *Museo di Storia Innaturale* (Museum of unnatural history), rug from wool fabric. Eight hanging rugs fashioned after Arab prayer carpets. Each has a brain portrayed at its center with certain parts colored, as if representing the specific areas related to the sin whose name is woven into the fabric.

Pierre Huyghe, *UUmwelt* (2012), freestanding LED panels. Depicts "mental images": subjects under an fMRI scanner were given verbal descriptions or shown pictures of living creatures, prehistoric tools, machines, codes, and artworks; the data were then processed via an AI neural network.

Michel Thiebaut de Schotten and Benedicte Batrancourt, *Lace Brain* (2012), diffusion images, Photoshop color filters and percolator filter. Abstract cerebral illustration serves as the official cover image for the Organization for Human Brain Mapping.

Marco Donnarumma and Dmitry Morozov, *The Escalation of Mind* (2012), experimental performance. Human brain served as a control voltage generator, monitoring in real time the cerebral outputs, emotional state, and facial expressions of an actor who quoted passages from Herman Hesse's novel *The Glass Bead Game*. The signals were used to generate an audiovisual environment.

Angela Palmer, *Self Portraits 4* (2012), sculpture engraved in sixteen sheets of glass, 23.6 × 11.5 × 5.4 in. 3-D views of MRI and CAT scans etched in layers of glass.

Ursula Damm, *Chromatographic Ballads* (2013), interactive installation. Visitors outfitted with EEG devices were invited to watch as their brain signals, equated in the exhibit with unconscious reactions, altered images of public spaces.

Jan Fabre, *Do We Feel with Our Brain and Think with Our Heart?* (2014), fifteen-minute video. Artist interviews Giacomo Rizzolatti while both wear EEG headsets.

Suzanne Dikker, with Marina Abramovic, *Measuring the Magic of Mutual Gaze* (2014). Experiment-like setup called for two participants with EEG headsets to engage in a mutual gaze for thirty minutes while their brain waves were recorded and visualized. Synchronization of the brain waves triggered a lightning animation.

Greg Dunn and Brian Edwards, *Self Reflected* (2014–2016), reflective microetching. Simulation of the electrical activity of the brain looking at art.

Laura Jade, *Cerebral Nebula* (2015), sculpture. Life-size human brain sculpture from MRI scan of thirty-five-year-old male, sliced into cross sections, laser cut in Perspex, and hand etched slice by slice with neural networks. Prototype for the *Brain Light Project*.

Jane Prophet, *Neuro memento mori* (2015), 3-D printed sculpture, polymer. Taken from MRI data of the artist while she meditated in the scanner. Inspired by a nineteenth-century wax model of a female head representing life and death.

Ellen Pearlman, *Noor: A Brain Opera* (2016), interactive immersive brain opera. Audiovisual performance of the story of Noor Inayat Khan, a covert wireless operator who infiltrated Vichy France.

Michela Martello, *Grandma's Mirror* (2017), acrylic, antique lace, marker or torsello, handmade vintage linen, 27 × 41 in. Juxtaposes a bright pink drawing of a brain with an antique lace featuring imagery from the ancient world.

Laura Jacobson, *Resonance Punctuated CXLIV* (2018), ceramic works. Stains, mounted with epoxy on painted wood, 12 × 9 × 1.5 in. Cross sections of a human MRI taken from the axial direction (i.e., horizontal to the upright body).

Lara Favaretto, *Thinking Head* (2019), poster art for the project Clandestine Talks. Ironically reprises the nineteenth-century phrenological enterprise by localizing functions such as animism and self-doubt on the skull. This served as a topic for the talks.

Pierre Huyghe, *Mental Images* (2020), LED panels. Mental images progressively reconstructed from brain activity via artificial neural networks.

# INDEX

Printed and bound by CPI Group (UK) Ltd, Croydon, CR0 4YY
13/10/2022
03154562-0002